D1208159

MOLECULAR ELECTROMAGNETISM

Molecular Electromagnetism

Alan Hinchliffe and Robert W. Munn
Department of Chemistry
University of Manchester Institute of Science and Technology

JOHN WILEY & SONS LIMITED
Chichester · New York · Brisbane · Toronto · Singapore

Library of Congress Cataloging in Publication Data:
Hinchliffe, Alan.
Molecular electromagnetism.
Includes bibliographies and index.
1. Electromagnetism. 2. Molecules. I. Munn, Robert W.
II. Title.
OC760.H55 1985 537 85–628

ISBN 0 471 10292 X (cloth)
ISBN 0 471 90721 9 (paper)

British Library Cataloguing in Publication Data:
Hinchliffe, Alan
Molecular electromagnetism.
1. Electromagnetism
I. Title II. Munn, Robert W.
537 QC760

ISBN 0 471 10292 X (cloth)
ISBN 0 471 90721 9 (paper)

Typeset by Mathematical Composition Setters Limited
7 Ivy Street, Salisbury, Wiltshire, UK
Printed by Biddles Ltd., Guildford, Surrey

Contents

Preface

Electromagnetic interactions pervade molecular science. Notwithstanding recent successes in obtaining a unification of the basic forces in nature, it remains convenient to distinguish four types: the gravitational, the electromagnetic, and the weak and strong nuclear forces. Of these, the gravitational force is important on a scale of distance ranging from the macroscopic to the astronomical and the nuclear forces on a sub-atomic scale, leaving the electromagnetic force to span the chemically important range from the atomic to the macroscopic. Thus Coulomb forces determine the physical interaction between nuclei and electrons, between polar molecules, between ions in solution and between colloidal particles. Much of our experimental knowledge of molecular systems is obtained by studying the way in which they respond to applied electric and magnetic fields, and most notably from their response to electromagnetic radiation in spectroscopic experiments.

A proper grounding in molecular science must therefore include an understanding of electromagnetism and electromagnetic properties. It is sometimes said that the whole of electromagnetism is merely the solution of Maxwell's equations under different conditions, and indeed many texts start at just that point. Whilst this approach may be satisfactory for mathematicians or electrical and electronic engineers, it is not the optimal one for physical scientists. Much physics is encapsulated within the Maxwell equations, and their postulation was in itself a highly significant achievement.

Books on electromagnetism abound for all levels and for many disciplines. Almost without exception these contain little that is explicitly molecular in content. Books on molecular electromagnetic properties are less numerous, particularly at the undergraduate level, where, however, many books give a passing mention to a few properties. The present book is designed to satisfy the resulting need for a self-contained and reasonably comprehensive account of electromagnetism and properties of molecular systems. It is based on a course we have given for many years to final year undergraduates in chemical physics at UMIST, and so should be most suitable for students at that level. The more descriptive parts could also prove accessible in the earlier undergraduate years, while the integrated treatment we provide should prove useful as a conversion or refresher course for graduates with a less complete or less organized background in this area who are beginning research in one of the interdisciplinary regions of molecular science such as polymer science or molecular electronics.

The book is divided into three parts. Part A deals with the basic principles of electromagnetism, starting from the fundamental experimental observations and building up to Maxwell's equations, from which the properties of electromagnetic waves are deduced and analysed, with a brief account of relativity as it relates to electromagnetism. This part is largely self-contained, but readers should have a prior acquaintance with descriptive electricity and magnetism. Vector notation and particularly vector calculus are extensively used because of their economy and power; whilst the reader is assumed to be familiar with vector algebra, the necessary calculus is summarized in an appendix. Readers should also be familiar with complex numbers.

Part B gives an account of the electromagnetic properties of molecules and assemblies of molecules from a classical phenomenological viewpoint, where molecules are treated as collections of charges with magnetic moments which can produce and respond to external fields. In this part, an appreciation of the role of the Boltzmann factor in determining thermal probabilities is required.

Part C goes on to treat the quantum-mechanical origins of electromagnetic behaviour. Readers are required to have an intermediate knowledge of quantum mechanics and its application to atoms and molecules, but the basic principles of quantum mechanics are reviewed both as revision and as a source of reference for later in the book. Two basic themes are covered in this part: the response of molecules to electromagnetic fields and to electromagnetic waves. The former leads to the calculation of the phenomenological coefficients introduced in Part B, while the latter leads to the calculation of spectroscopic coefficients, both being based on a detailed treatment of change of quantum states. Our treatment of spectroscopy is designed to establish the underlying electromagnetic and quantum-mechanical principles, which specialized textbooks on spectroscopy often gloss over in order to concentrate on details of spectroscopic measurement and interpretation which we omit completely.

Our thanks are due to Professor D. W. J. Cruickshank, FRS, who originated the chemical physics course at UMIST and with it the idea of electromagnetism as a major constituent. We must therefore share some of the credit for this book with him, while retaining all responsibility for errors and omissions ourselves. We also thank a succession of editors at John Wiley and Sons Limited for their continuing support and encouragement.

January 1985

Alan Hinchliffe
Robert W. Munn

Part A
Basic electromagnetism

Chapter 1

Electrostatics

1.1 Coulomb's law

The whole subject of electromagnetism is bound up with the idea of electric charge. Therefore we begin the discussion with an investigation of the fields due to stationary electric point charges in free space. In Chapter 2 we show how the fields are modified by the presence of matter, and in later chapters examine the fields associated with moving electric charges.

The existence of electric charge was known to the ancient Greeks who found that a rod of amber when rubbed with fur would attract small pieces of paper. No great advances were made, however, until around 1600 when Gilbert began a detailed study of the kinds of materials that would behave like amber. The general experimental facts about electric charge can be summarized as follows:

(1) Electric charge can exist in matter.
(2) There are two kinds of charge, called by convention *positive* and *negative* charge. We are rarely aware of the existence of electric charge in nature because positive and negative charges are usually exactly balanced. The *particles* responsible for electric charge in bodies are of course electrons and protons. These are certainly *not point charges* in the true mathematical sense, but classical electromagnetism is not concerned with phenomena occurring on the length scale of such particles, as we shall see in some detail in Chapters 2 and 4. We therefore begin the discussion with an investigation into the electrostatic forces between point charges. In later chapters we show how quantum mechanics is needed when discussing the electric and magnetic properties of atomic particles.
(3) It is known experimentally that the forces between point charges act along their lines of centres.
(4) It is known experimentally that the magnitude of the force is proportional to r^{-n}, where r is the scalar distance between the point charges. Cavendish showed that $n = 2.00 \pm 0.02$ but it is now known that n is equal to 2 within 1 part in 10^9 and there is no reason to believe that n is not exactly equal to 2. There are four kinds of force in nature: electrostatic and gravitational forces which are both inverse square laws, together with strong and weak nuclear forces. Of the former pair the electrostatic force is usually the

dominant one when both are present, but the latter pair of nuclear forces are not well understood experimentally. They certainly fall off faster than r^{-2} and are completely unimportant outside the radius of a nucleus (10^{-14} m), but whilst the 1979 Nobel physics prize was won for a partial unification of the fields involved, a unified field theory is still absent.

(5) It is known experimentally that the electrostatic force is proportional to $q_a q_b$, where q_a and q_b are point charges. The force is attractive if $q_a q_b$ is negative, repulsive if the product is positive.

Coulomb's law is a statement of the above experimental results:

$$\boldsymbol{F}_{b,a} = \frac{1}{4\pi\epsilon_0} q_b q_a \frac{\boldsymbol{r}_{b,a}}{r_{b,a}^3} \tag{1.1}$$

where $\boldsymbol{F}_{b,a}$ is the force exerted *by* q_b *on* q_a and $\boldsymbol{r}_{b,a} = \boldsymbol{r}_a - \boldsymbol{r}_b$ is a vector drawn from q_b to q_a. Equation (1.1) satisfies Newton's third law since the force exerted on q_a by q_b is the negative of the force exerted on q_b by q_a, the vector $\boldsymbol{r}_{a,b} = -\boldsymbol{r}_{b,a}$ ensuring the change of sign in eq. (1.1). The factor 4π in eq. (1.1) implies a rationalized system of units and in SI the proportionality constant ϵ_0 is the *permittivity of free space*, an experimentally determined quantity with the value

$$\epsilon_0 = 8.854\,187\,82 \times 10^{-12}\ \mathrm{F\,m^{-1}}. \tag{1.2}$$

Coulomb's law, eq. (1.1), can be written in different ways to emphasize different points: thus to emphasize the inverse square nature of eq. (1.1)

$$\boldsymbol{F}_{b,a} = \frac{q_b q_a}{4\pi\epsilon_0} \frac{\hat{\boldsymbol{r}}_{b,a}}{r_{b,a}^2} \tag{1.3}$$

where $\hat{\boldsymbol{r}}_{b,a}$ is a unit vector pointing from q_b to q_a, and if it is desired to express the force in terms of the position vectors of q_a and q_b expressed in terms of the common origin of coordinates shown in Fig. 1.1,

$$\boldsymbol{F}_{b,a} = \frac{q_b q_a}{4\pi\epsilon_0} \frac{\boldsymbol{r}_a - \boldsymbol{r}_b}{|\boldsymbol{r}_a - \boldsymbol{r}_b|^3}. \tag{1.4}$$

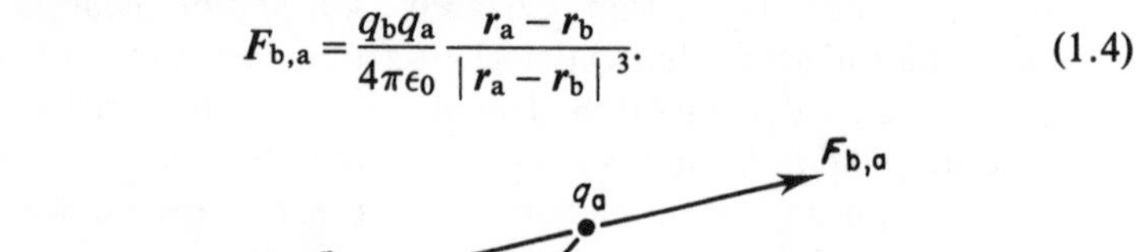

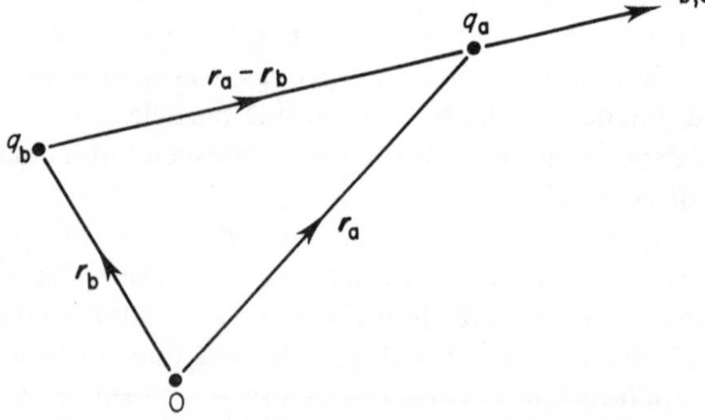

Fig. 1.1 Coordinates used in Coulomb's law, eq. (1.4)

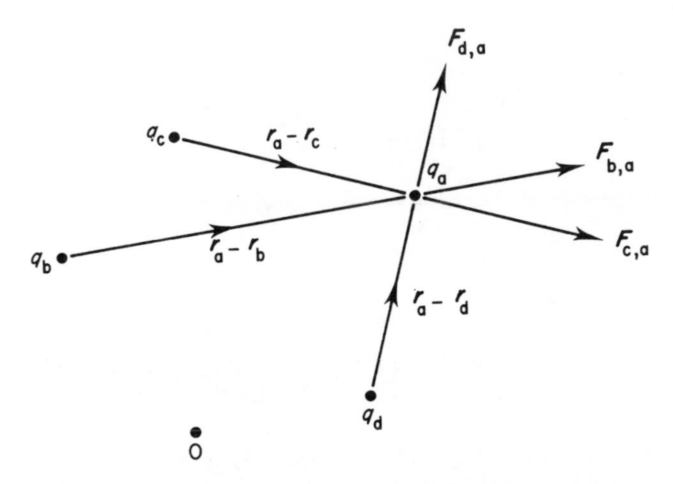

Fig. 1.2 The superposition principle; the total force on q_a is the sum of the forces due individually to the other point charges

1.2 Superposition principle

Electrostatic forces between pairs of point charges are unaltered by the addition of further point charges. Thus the force $\boldsymbol{F}_{b,a}$ of eq. (1.1) is unchanged when extra point charges $q_c, q_d, \ldots, q_n$ are added, as in Fig. 1.2. Such forces are said to be *pair-wise additive* and the total force $\boldsymbol{F}_a$ on point charge q_a due to point charges $q_b, q_c, \ldots, q_n$ is

$$\boldsymbol{F}_a = \sum_p \boldsymbol{F}_{p,a} = \frac{q_a}{4\pi\epsilon_0}\sum_p \frac{q_p(\boldsymbol{r}_a - \boldsymbol{r}_p)}{|\boldsymbol{r}_a - \boldsymbol{r}_p|^3}, \tag{1.5}$$

where $\boldsymbol{r}_p$ is the position vector of point charge q_p relative to the coordinate origin of Fig. 1.2. This additive property of electrostatic forces between point charges is known as the *superposition principle.*

1.3 Continuous charge distributions

In general, attention is more usually focused on *continuous* distributions of electric charge rather than on arrays of point charges. For such continuous charge distributions the summation of eq. (1.5) must clearly be replaced with an integral. We will use the following symbols throughout the text for charge distributions:

$\varrho(\boldsymbol{r}) \equiv \mathrm{d}q/\mathrm{d}\tau$, volume charge density
$\sigma(\boldsymbol{r}) \equiv \mathrm{d}q/\mathrm{d}A$, surface charge density
$\mu(\boldsymbol{r}) \equiv \mathrm{d}q/\mathrm{d}l$, line charge density

where $\mathrm{d}\tau$, $\mathrm{d}A$ and $\mathrm{d}l$ are respectively volume, area and line elements. Thus

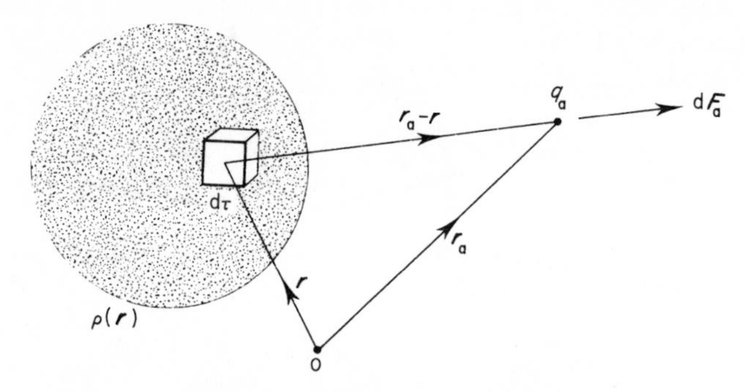

Fig. 1.3 The force on a point charge q_a due to a continuous charge distribution $\varrho(r)$

$dq = \varrho(r)\, d\tau$, etc., as in Fig. 1.3. and eq. (1.4) becomes

$$dF_a = \frac{q_a}{4\pi\epsilon_0}\,\varrho(r)\,\frac{(r_a - r)}{|r_a - r|^3}\, d\tau. \tag{1.6}$$

The total force on q_a, F_a, is found by integrating eq. (1.6):

$$F_a = \frac{q_a}{4\pi\epsilon_0}\int \varrho(r)\,\frac{(r_a - r)}{|r_a - r|^3}\, d\tau \tag{1.7}$$

(where we have adopted the notation of Appendix A for a volume integral: in Cartesian coordinates eq. (1.7) would be

$$F_a = \frac{q_a}{4\pi\epsilon_0}\iiint \varrho(x, y, z)\,\frac{(r_a - r)}{|r_a - r|^3}\, dx\, dy\, dz).$$

Example

Figure 1.4 shows a point charge q_a lying along on an axis of a uniformly charged rod of length L. The total force exerted by the rod on q_a clearly points in the direction shown, and eq. (1.7) yields

$$dF = \frac{q_a}{4\pi\epsilon_0}\,\frac{dq}{r^2}$$

and so by integration

$$F = q_0\mu L/4\pi\epsilon_0\, L(a + L).$$

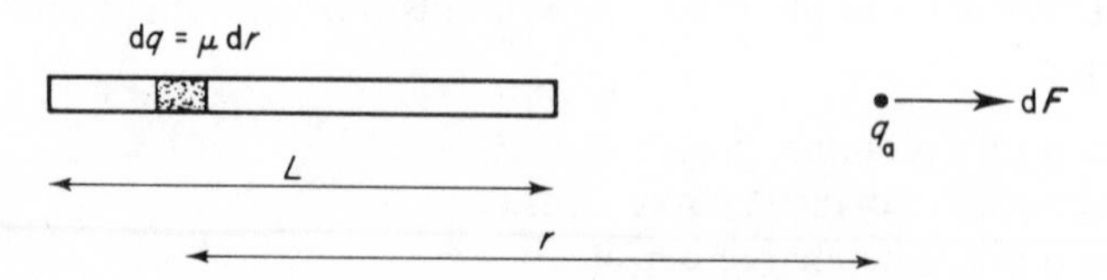

Fig. 1.4 Figure for worked example

The generalization of eq. (1.7) to cover the case of the force between two interacting continuous charge distributions $\varrho_a(\boldsymbol{r}_a)$ and $\varrho_b(\boldsymbol{r}_b)$ is

$$\boldsymbol{F}_a = \frac{1}{4\pi\epsilon_0} \iint \varrho_a(\boldsymbol{r}_a)\, \varrho_b(\boldsymbol{r}_b) \frac{(\boldsymbol{r}_a - \boldsymbol{r}_b)}{|\boldsymbol{r}_a - \boldsymbol{r}_b|^3} \mathrm{d}\tau_a \mathrm{d}\tau_b \tag{1.8}$$

where the notation is shorthand for a six-dimensional integral. The superposition principle however, does not necessarily apply when charge distributions are concerned: a charge distribution $\varrho_a(\boldsymbol{r}_a)$ could be *polarized* by the successive addition of point charges q_b, q_c, ..., q_n and so the total force on $\varrho_a(\boldsymbol{r}_a)$ would not be the sum of the forces $\boldsymbol{F}_{i,a}$ on $\varrho_a(\boldsymbol{r}_a)$ due to the point charges in the absence of the other point charges. We assume that point charges are not polarizable.

1.4 Electric fields

The force experienced by the point charge q_a in the presence of point charges $q_b, q_c, \ldots, q_n$ can be interpreted as showing the interaction of q_a with the *electrostatic field* $\boldsymbol{E}$ *produced by* q_b, q_c, ..., q_n. The equation

$$\boldsymbol{F}_a = q_a \boldsymbol{E} \tag{1.9}$$

gives a formal definition of the electrostatic field. Comparing eqs (1.5) and (1.9) shows that the electrostatic field at the position of q_a is

$$\boldsymbol{E}(\boldsymbol{r}_a) = \frac{1}{4\pi\epsilon_0} \sum q_i \frac{(\boldsymbol{r}_a - \boldsymbol{r}_i)}{|\boldsymbol{r}_a - \boldsymbol{r}_i|^3} \tag{1.10}$$

and it is important to note that eq. (1.10) contains no reference to the point charge q_a. $\boldsymbol{E}(\boldsymbol{r})$ is clearly a vector field that exists at all points in space and does not depend on the presence of the *test charge* q_a, so it is usual to drop all reference to 'a' and rewrite eq. (1.10) as

$$\boldsymbol{E}(\boldsymbol{r}) = \frac{1}{4\pi\epsilon_0} \sum q_i \frac{(\boldsymbol{r} - \boldsymbol{r}_i)}{|\boldsymbol{r} - \boldsymbol{r}_i|^3} \tag{1.11}$$

or, taking the field point $\boldsymbol{r}$ as origin

$$\boldsymbol{E}(\boldsymbol{r}) = \frac{1}{4\pi\epsilon_0} \sum q_i \frac{\boldsymbol{r}_i}{r_i^3}. \tag{1.12}$$

Because of the problem mentioned above concerning the polarization of charge distributions, the *formal* definition of $\boldsymbol{E}$ afforded by eq. (1.9) should be generalized to

$$\boldsymbol{E} = \operatorname*{Lt}_{\delta q \to 0} \frac{\delta \mathbf{F}_a}{\delta q_a} \tag{1.13}$$

but eq. (1.12) or

$$\boldsymbol{E} = \frac{1}{4\pi\epsilon_0} \int \frac{\boldsymbol{r}}{r^3} \mathrm{d}q \tag{1.14}$$

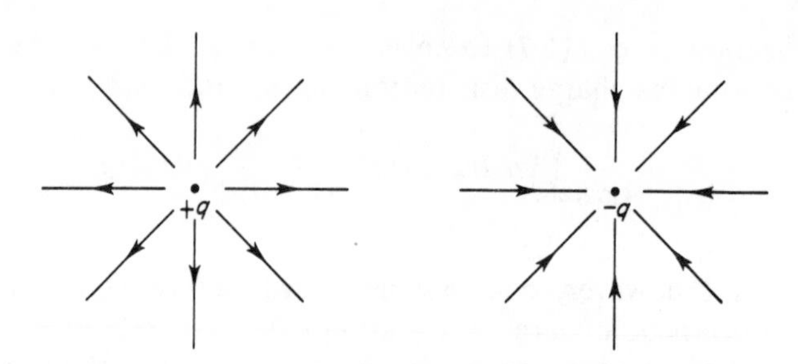

Fig. 1.5 Electric flux lines for positive and negative point charges. Flux lines always start at positive charges and terminate at negative charges

which corresponds to eq. (1.12) are true since they contain no reference to the test charge q_a.

Like all vector fields, electric fields are often represented pictorially by *flux lines* as in Fig. 1.5, the volume density of lines representing the electric field strength. In some ways, however, it can be misleading to use such a concept because unlike for example the hydrodynamic case, nothing is *flowing* in electro*statics* (but see Chapter 6). With this proviso the physical picture conveyed by Fig. 1.5 can be very useful. An important point to note is that electric flux lines *always* have beginnings and ends; they start at positive charges and terminate at negative, and this behaviour should be contrasted with the behaviour of magnetic flux lines to be discussed in Chapter 3.

1.5 Work

In this section we show that the work done in moving an electric charge from point A to point B in an electrostatic field is independent of the path taken between A and B. The calculation depends on the idea of line integrals (Appendix A), and the conclusion is that electrostatic fields, which are *vector* fields, can be defined by a *scalar* electrostatic potential.

Consider a test charge q_0 that can be moved in an electrostatic field $\boldsymbol{E}(\boldsymbol{r})$. The work done in moving q_0 in the field is given by the force on q_0 multiplied by the distance moved in the direction of the field

$$\mathrm{d}w = -\boldsymbol{F}\cdot\mathrm{d}\boldsymbol{l} \tag{1.15}$$

where the negative sign is needed in order to obtain the work done *against* the field. Thus the work done moving q_0 from point A to point B along a path $\mathrm{P_I}$ is

$$\begin{aligned} w_{\mathrm{I}} &= -\int_{\mathrm{P_I}} \boldsymbol{F}\cdot\mathrm{d}\boldsymbol{l} \\ &= -q_0\int_{\mathrm{P_I}} \boldsymbol{E}\cdot\mathrm{d}\boldsymbol{l}. \end{aligned} \tag{1.16}$$

In order to show that E is a *conservative* field we need to show that the work done in moving q_0 between A and B is independent of path. For simplicity, and without loss of generality because of the superposition principle, the electric field E can be treated as though it is produced by a single point charge Q. Thus from eq. (1.12)

$$E = \frac{Q}{4\pi\epsilon_0}\frac{r}{r^3} \tag{1.17}$$

and from Fig. 1.6 we see that

$$\begin{aligned} \mathrm{d}w &= -q_0 E \cdot \mathrm{d}l \\ &= -q_0 E \;\mathrm{d}l \cos\theta \\ &= -q_0 E \;\mathrm{d}r \\ &= -q_0 \frac{Q}{4\pi\epsilon_0}\frac{r}{r^3}\,dr \end{aligned}$$

so that the *total* work done in moving q_0 from A to B is

$$w_{A\to B} = -\int_A^B \frac{q_0 Q}{4\pi\epsilon_0}\frac{\mathrm{d.}}{r^2}$$

whence

$$w_{A\to B} = \frac{q_0 Q}{4\pi\epsilon_0}\left\{\frac{1}{r_B} - \frac{1}{r_A}\right\} \tag{1.18}$$

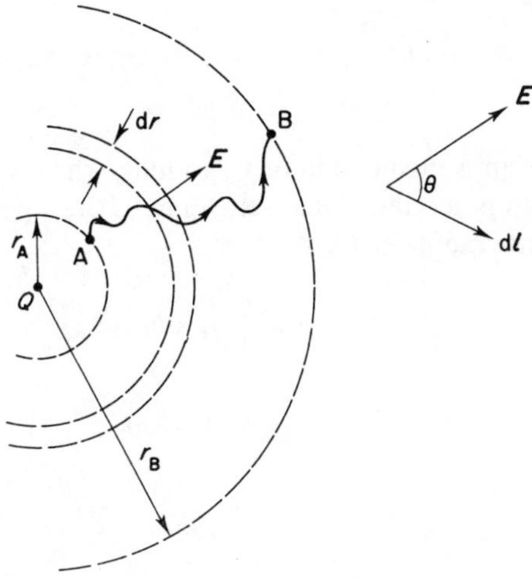

Fig. 1.6 Geometric construct used to demonstrate that electrostatic fields are conservative. The test charge q_0 moves from position A to position B by the route indicated, in a field provided by the point charge Q

and the important point regarding eq. (1.18) is that $w_{A\to B}$ contains no reference whatever to the path linking A and B. For electrostatic fields this is always true and so we conclude (Appendix A) that *electrostatic fields are conservative fields.* Thus the following general properties of an arbitrary conservative field $\boldsymbol{A}$ apply to the electrostatic field:

(1) If $\boldsymbol{A}$ is a conservative vector field then

$$\oint \boldsymbol{A}\cdot d\boldsymbol{l} = 0$$

for any closed path.

(2) If $\boldsymbol{A}$ is a conservative vector field then

$$\nabla \times \boldsymbol{A} = \boldsymbol{0}$$

at all points for which $\boldsymbol{A}$ is defined.

(3) If $\boldsymbol{A}$ is a conservative *vector* field, it is possible to find a *scalar* field V such that

$$\boldsymbol{A} = \nabla V$$

This latter scalar field has a very special significance and we now show how the electrostatic potential $V(\boldsymbol{r})$ can be found.

1.6 Electrostatic potential

The *electrostatic scalar potential* $V(\boldsymbol{r})$ is, according to Appendix A, defined as

$$\boldsymbol{E} = -\nabla V \tag{1.19}$$

where the negative sign is included in order to make the vector field $\boldsymbol{E}(\boldsymbol{r})$ point towards a decrease in potential. Thus, returning to the algebra of Section 1.5, the work done in moving q_0 from A to B is

$$\begin{aligned} w &= -q_0 \int_A^B \boldsymbol{E}\cdot d\boldsymbol{l} \\ &= q_0 \int_A^B \nabla V\cdot d\boldsymbol{l} \\ &= q_0 \int_A^B dV \end{aligned}$$

which can be written

$$w_{A\to B} = q_0(V_B - V_A). \tag{1.20}$$

It is important to realize that the potential $V(\boldsymbol{r})$ is undetermined to within a constant of integration, because of its definition by eq. (1.19) as a *differential.* By convention, the constant of integration is usually fixed by setting $V=0$ at $r=\infty$, so the potential at point $\boldsymbol{r}$ is calculated as the work done in bring test-charge q_0 from infinity to field point $\boldsymbol{r}$.

$$w = q_0 V(\boldsymbol{r}) - q_0 V(r = \infty) \tag{1.21}$$

and comparing eqs (1.21) and (1.18) it is clear that the potential due to a point charge q is

$$V(\boldsymbol{r}) = \frac{q}{4\pi\epsilon_0 r} \tag{1.22}$$

at field point $\boldsymbol{r}$ taken as origin of the coordinate system.

Potentials are measured in volts (V) in SI, and the sign of the potential V is the same as the sign of q. Pair-wise additivity considerations apply as in Section 1.2 and the potential due to a continuous distribution of charge $\varrho(\boldsymbol{r})$ is given by the three-dimensional integral

$$V(\boldsymbol{r}) = \int \frac{\mathrm{d}q}{4\pi\epsilon_0 r} \equiv \int \frac{\varrho(\boldsymbol{r})\,\mathrm{d}\tau}{4\pi\epsilon_0 r}. \tag{1.23}$$

Electric fields are often represented as flux-lines, and it is usual to draw electrostatic potentials as contour diagrams, since such potentials are scalars whose value at a given distance from a charge distribution is important.

We now illustrate how $\boldsymbol{E}$ and V may be found by direct application of eqs (1.14) and (1.23). As a general practical guide, unless the direction of $\boldsymbol{E}$ is clear (e.g. by reason of symmetry), it is easier to find an expression for $V(\boldsymbol{r})$, which is a *scalar* field, and hence find $\boldsymbol{E}(\boldsymbol{r})$ from $\boldsymbol{E} = -\nabla V$. The first two examples are cases where the direction of $\boldsymbol{E}$ is obvious, and $\boldsymbol{E}$ can be calculated by the direct application of eq. (1.23). In the third example, it is arguably easier to proceed *via* the scalar potential.

Field due to long uniformly charged wire

If the charge density is uniform along the wire, then the parallel component of $\boldsymbol{E}$ due to the wire element $\mathrm{d}l$ in Fig. 1.7 corresponding to an angle θ will exactly cancel that due to the element corresponding to $-\theta$. Hence the direction of $\boldsymbol{E}$ must be $\hat{\boldsymbol{n}}$ as indicated, and assuming for the moment that the wire has 'infinite' length

$$\boldsymbol{E}_\perp = \hat{\boldsymbol{n}} \int_{\theta=-\pi/2}^{\theta=+\pi/2} \frac{\cos\theta}{4\pi\epsilon_0 r^2}\,\mathrm{d}q,$$

where the term $\cos\theta$ gives the perpendicular component of the field. Expressing the integrand in terms of the variable θ and the constant a where

$$l = a\tan\theta \qquad a = r\cos\theta$$
$$\mathrm{d}l = a\sec^2\theta\,\mathrm{d}\theta \qquad \mathrm{d}q = \mu\,\mathrm{d}l$$

we find

$$\boldsymbol{E}_\perp = \hat{\boldsymbol{n}}\frac{\mu}{3\pi\epsilon_0 a}. \tag{1.24}$$

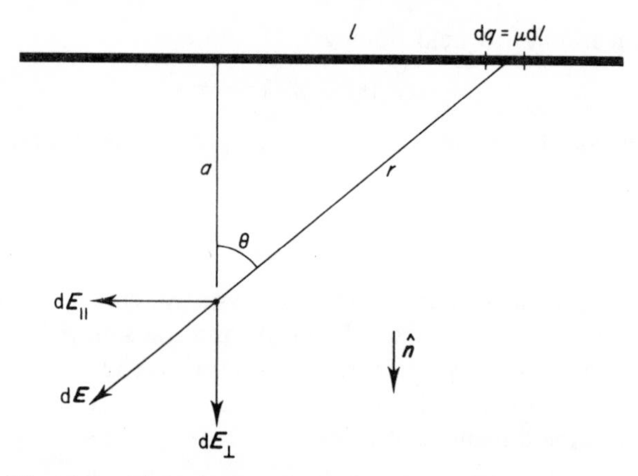

Fig. 1.7 Calculation from first principles of the electrostatic field due to a long, uniformly charged wire

Field due to hollow charged sphere

The shaded area of Fig. 1.8 represents a band of surface corresponding to angles θ and $\theta + \mathrm{d}\theta$. By symmetry the resultant $\boldsymbol{E}$ must be normal to the sphere surface, all tangential components cancelling. The surface area $\mathrm{d}S$ of this band is

$$\mathrm{d}S = 2\pi\ r \sin\theta\ r\mathrm{d}\theta$$

so if the uniform surface charge density is σ,

$$\mathrm{d}q = 2\pi r^2 \sin\theta\ \sigma\ \mathrm{d}\theta.$$

Thus

$$\boldsymbol{E}_n = \hat{\boldsymbol{n}} \frac{1}{4\pi\epsilon_0} \int_{\theta=0}^{\theta=2\pi} 2\pi r^2 \sin\theta\ \sigma \frac{\cos\phi}{x^2}\,\mathrm{d}\theta$$

and using the cosine rule

$$\cos\phi = \frac{x^2 + R^2 - r^2}{2xR}; \qquad \cos\theta = \frac{r^2 + R^2 - x^2}{2rR}$$

to express the integrand in terms of the variable x and the constants r and R, we find

$$\boldsymbol{E}_n = \hat{\boldsymbol{n}} \frac{Q}{4\pi\epsilon_0 R^2}, \tag{1.25}$$

where $Q = 4\pi r^2\sigma$ is the total charge on the sphere. This result is formally identical to the electric field of a point charge q at the centre of the field. It is an experimental fact that the charge on a hollow conductor resides on the surface of the conductor, and the electric field inside such a conductor is zero.

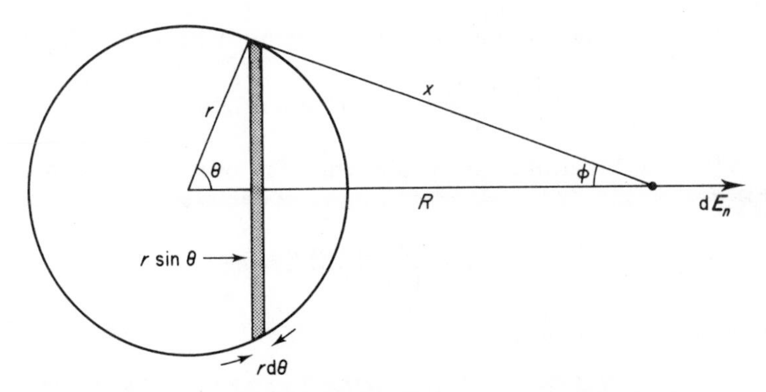

Fig. 1.8 Calculation from first principles of the electrostatic field due to a hollow charged sphere

Potential and field due to electric dipole

In its simplest form, an electric dipole consists of equal and opposite point electric charges $\pm q$ separated by scalar distance d. The magnitude of the dipole $\boldsymbol{p}$ is then qd.

From eq. (1.22) for the potential due to a point charge

$$V = \frac{q}{4\pi\epsilon_0}\left\{\frac{1}{r_+} - \frac{1}{r_-}\right\}$$

and from the cosine rule

$$r_+ = r^2 + \frac{1}{4}d^2 - dr\cos\theta$$

so that

$$\frac{1}{r_+} = \frac{1}{r}\left\{1 + \frac{d^2}{4r^2} - \frac{d}{r}\cos\theta\right\}^{-\frac{1}{2}}$$

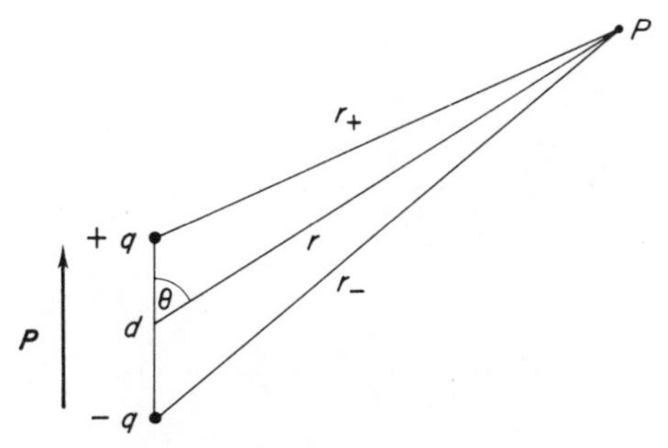

Fig. 1.9 Calculation of the electrostatic field and potential due to a small, simple dipole

which can be expanded by the binomial theorem to give

$$\frac{1}{r_+} = \frac{1}{r} - \frac{d\cos\theta}{2r^2} + \text{higher order terms.}$$

Assuming that the field point is sufficiently far away from the dipole such that $r \gg d$,

$$V = \frac{q}{4\pi\epsilon_0}\frac{d\cos\theta}{r^2}$$

or, in terms of the dipole $\boldsymbol{p}$,

$$V = \frac{\boldsymbol{p}\cdot\boldsymbol{r}}{4\pi\epsilon_0 r^3}. \tag{1.26}$$

Equation (1.26) is true generally for an electric dipole $\boldsymbol{p}$ irrespective of whether $\boldsymbol{p}$ arises from a pair of charges or from a more complex distribution of electric charge when

$$\boldsymbol{p} = \sum q_i \boldsymbol{r}_i + \int \boldsymbol{r}\varrho(\boldsymbol{r})\,d\tau.$$

The electric field E can be found from $\boldsymbol{E} = -\nabla V$. Thus

$$\boldsymbol{E} = -\nabla\left[\frac{\boldsymbol{p}\cdot\boldsymbol{r}}{4\pi\epsilon_0 r^3}\right]$$

$$= -\frac{1}{4\pi\epsilon_0}\left\{\frac{1}{r^3}\nabla(\boldsymbol{p}\cdot\boldsymbol{r}) + (\boldsymbol{p}\cdot\boldsymbol{r})\nabla\left(\frac{1}{r}\right)\right\}.$$

Now it is easily shown (see Examples 1.2 and 1.3) that

$$\nabla(\boldsymbol{p}\cdot\boldsymbol{r}) = \boldsymbol{p}$$

and

$$\nabla\left(\frac{1}{r^3}\right) = -\frac{3\boldsymbol{r}}{r^5}.$$

Hence

$$\boldsymbol{E} = -\frac{1}{4\pi\epsilon_0}\left\{\frac{\boldsymbol{p}}{r^3} - 3\frac{(\boldsymbol{p}\cdot\boldsymbol{r})}{r^5}\boldsymbol{r}\right\}$$

or

$$\boldsymbol{E} = \frac{1}{4\pi\epsilon_0}\left\{\frac{3(\boldsymbol{p}\cdot\boldsymbol{r})\boldsymbol{r} - r^2\boldsymbol{p}}{r^5}\right\}. \tag{1.27}$$

The electric field due to a dipole thus falls off much faster (r^{-3} versus r^{-2}) than that due to a point charge (or *monopole*), because in the dipole case the equal and opposite charges help cancel out each other's effects.

1.7 Gauss' theorem

The electric field due to a point charge q is

$$\boldsymbol{E} = \frac{q}{4\pi\epsilon_0}\frac{\boldsymbol{r}}{r^3}$$

and the field has spherical symmetry. The electric flux passing through a sphere of radius R centred at the point charge is

$$\begin{aligned}\text{Flux} &= \text{surface area} \times \text{component of } E \text{ perpendicular to surface} \\ &= 4\pi R^2 \times E \\ &= q/\epsilon_0\end{aligned} \tag{1.28}$$

independent of R. Expression (1.28) is only correct because $\boldsymbol{E}$ is constant at all points on the sphere and everywhere perpendicular to the sphere. In a general case where an arbitrary surface S encloses an arbitrary charge distribution, the direction of $\boldsymbol{E}$ and its magnitude will be different at different points on S and eq. (1.28) will need to be generalized. From Appendix A, the flux through such a surface is given by a *surface integral.*

$$\int \boldsymbol{E}\cdot \mathrm{d}\boldsymbol{S}$$

and Gauss' theorem states that the total flux through *any closed* surface S enclosing charge q is q/ϵ_0:

$$\oint \boldsymbol{E}\cdot \mathrm{d}\boldsymbol{S} = q/\epsilon_0. \tag{1.29}$$

If no charge is enclosed by the surface S then

$$\oint \boldsymbol{E}\cdot \mathrm{d}\boldsymbol{S} = 0.$$

For simplicity and without loss of generality (because of the superposition principle), we consider the case where $\boldsymbol{E}$ is due to a single point charge q and do not for the moment consider surfaces shaped so that the electric flux lines can cut the surface more than once.

To find the flux through S, the surface is divided into surface elements $\mathrm{d}\boldsymbol{S}$, and the flux through $\mathrm{d}\boldsymbol{S}$ is

$$\boldsymbol{E}\cdot \mathrm{d}\boldsymbol{S}. \tag{1.30}$$

Figure 1.10 illustrates that the projected area of $\mathrm{d}S$ perpendicular to $\boldsymbol{r}$ is

$$\mathrm{d}S_\mathrm{p} = r \sin\theta \,\mathrm{d}\phi\, r\mathrm{d}\theta$$

so the flux through $\mathrm{d}S$ is

$$\frac{q}{4\pi\epsilon_0}\frac{1}{r^2} r \sin\theta\,\mathrm{d}\phi\, r\mathrm{d}\theta$$

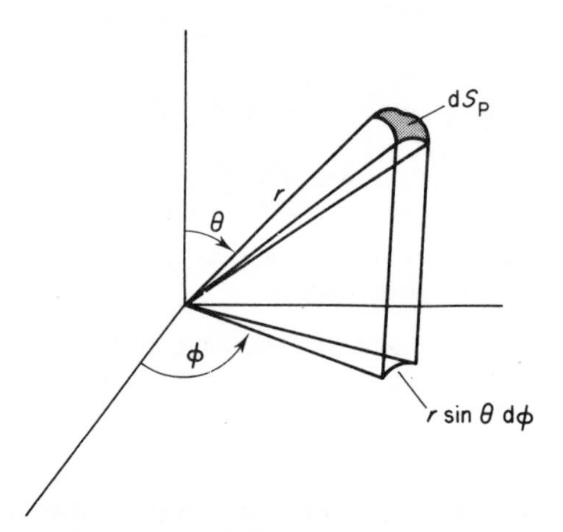

Fig. 1.10 Surface element dS_p for use in the proof of Gauss' theorem. dS_p is the projected area of surface element dS perpendicular to r

and the total flux through the closed surface S is

$$\oint \boldsymbol{E} \cdot d\boldsymbol{S} = \frac{q}{4\pi\epsilon_0} \int_0^{\pi} \sin\theta \, d\theta \int_0^{2\pi} d\phi = \frac{q}{\epsilon_0}$$

which proves Gauss' theorem. Equation (1.29) is the *integral* form of Gauss' theorem. It is valid in general whatever the shape of the *closed* surface S (which is usually referred to as a *gaussian surface*) and the flux leaving S makes a positive contribution to the surface integral $\oint \boldsymbol{E} \cdot d\boldsymbol{S}$ whilst the flux entering S makes a negative contribution. If the surface S encloses no charge, $\oint \boldsymbol{E} \cdot d\boldsymbol{S} = 0$ and the net flux through the surface is zero: as much electric flux enters as leaves. Gauss' theorem, which owes its validity to the inverse square law of electrostatic force, offers an elegant way of finding the electric field for charge distributions which have a high degree of symmetry. For a volume distribution of charge, eq. (1.29) becomes

$$\oint_S \boldsymbol{E} \cdot d\boldsymbol{S} = \frac{1}{\epsilon_0} \int_V \varrho(\boldsymbol{r}) d\tau \tag{1.31}$$

where the closed surface S encloses volume V. The choice of gaussian surface is arbitrary, but the principal problem is that of finding one that is suitable for the easy evaluation of the relevant surface integral: for simplicity it is necessary to find surfaces over which $\boldsymbol{E}$ is either parallel or perpendicular to S, over which $\boldsymbol{E}$ has constant magnitude, etc., for otherwise extraction of $\boldsymbol{E}$ from $\oint \boldsymbol{E} \cdot d\boldsymbol{S}$ will not be possible. Two examples illustrate how Gauss' theorem can be used in favourable cases to find $\boldsymbol{E}$.

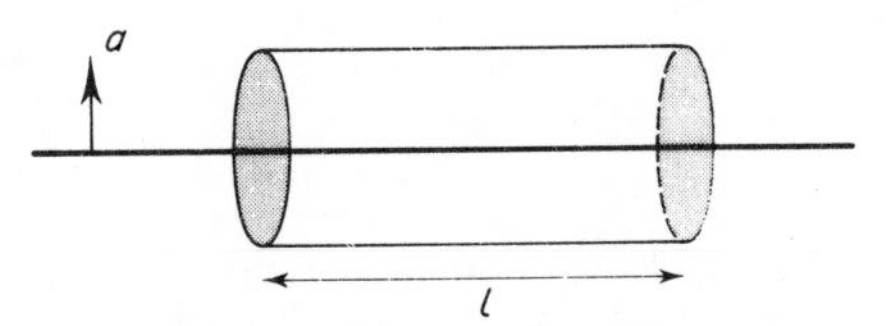

Fig. 1.11 Gaussian surface for the calculation of the electrostatic field due to a long, uniformly charged wire

Electric field due to long, uniformly charged wire

In Section 1.6 we found the field from first principles. An obvious choice for the gaussian surface is shown in Fig. 1.11, a cylinder concentric with the wire. With the assumption of infinite length, so that the electric flux is everywhere perpendicular to the wire, there is no flux through the circular ends of the cylinder. Thus

$$\oint \boldsymbol{E}\cdot d\boldsymbol{S} = \int_{\text{ends}} \boldsymbol{E}\cdot d\boldsymbol{S} + \int_{\text{sides}} \boldsymbol{E}\cdot d\boldsymbol{S} \tag{1.32}$$

and the first integral on the right-hand side of eq. (1.32) is zero. The field through the closed sides is given by

$$\int_{\text{sides}} \boldsymbol{E}\cdot d\boldsymbol{S} = E_n \times 2\pi a l.$$

Thus

$$E_n 2\pi a l = l\mu/\epsilon_0.$$

$$E_n = \frac{\mu}{2\pi\epsilon_0 a}$$

which is the result arrived at in Section 1.6, but by a shorter route.

Electric field due to thin infinite, uniform sheet of charge

Figure 1.12 shows a thin, infinite sheet of charge with uniform density σ. One possible suitable choice of gaussian surface is indicated: again by symmetry the field cannot have components parallel to the sheet and so the flux through the *sides* of the surface illustrated is zero. The shape of the gaussian surface is immaterial in this example, so long as the ends are parallel to the sheet. Hence

$$\oint \boldsymbol{E}\cdot d\boldsymbol{S} = 2EA = A\sigma/\epsilon_0$$

and so

$$E = \frac{\sigma}{2\epsilon_0}. \tag{1.33}$$

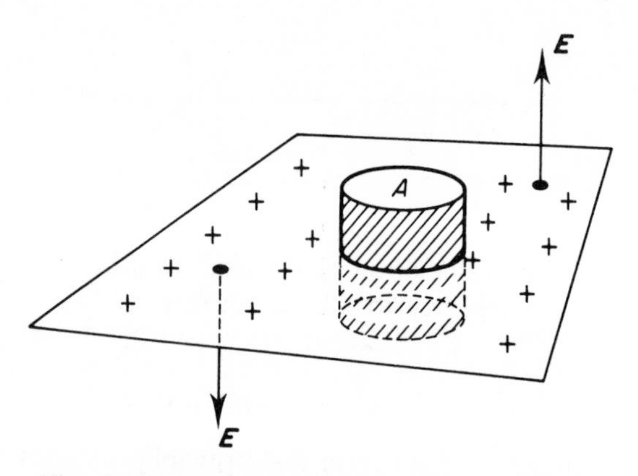

Fig. 1.12 Gaussian surface for calculation of the electrostatic field due to an infinite sheet of charge. The sheet is taken to have zero thickness

1.8 Differential form of Gauss' theorem

Gauss' theorem, eq. (1.29), is only practically useful when the charge distribution has a high degree of symmetry. We now show how the divergence theorem of Appendix A can be used to turn the integral eq. (1.29) into a differential equation relating the electric field $\boldsymbol{E}(\boldsymbol{r})$ at point $\boldsymbol{r}$ to the local charge density $\varrho(\boldsymbol{r})$. Finally, in view of the importance of the electrostatic potential in such calculations, we show how this differential form of Gauss' theorem can be recast in terms of the scalar potential at the point, $V(\boldsymbol{r})$.

Starting from the eq. (1.29)

$$\oint \boldsymbol{E}\cdot \mathrm{d}\boldsymbol{S} = \frac{1}{\epsilon_0}\int \varrho(\boldsymbol{r})\,\mathrm{d}\tau \tag{1.34}$$

we note that the divergence theorem can be used to relate the volume integral on the right-hand side of eq. (1.34) to a surface integral

$$\oint \boldsymbol{E}\cdot \mathrm{d}\boldsymbol{S} = \int_V \boldsymbol{\nabla}\cdot\boldsymbol{E}\,\mathrm{d}\tau$$

where the surface S encloses the volume V. Hence

$$\int_V \boldsymbol{\nabla}\cdot\boldsymbol{E}\,\mathrm{d}\tau = \frac{1}{\epsilon_0}\int \varrho(\boldsymbol{r})\,\mathrm{d}\tau \tag{1.35}$$

and because eq. (1.35) is true for any volume however small, the integrands must be equal, i.e.

$$\boldsymbol{\nabla}\cdot\boldsymbol{E} = \varrho(\boldsymbol{r})/\epsilon_0 \tag{1.36}$$

or

$$\frac{\partial E_x}{\partial x}+\frac{\partial E_y}{\partial y}+\frac{\partial E_z}{\partial z}=\varrho(\boldsymbol{r})/\epsilon_0. \tag{1.37}$$

Equation (1.37) is the differential form of Gauss' theorem: it describes the variation of $\boldsymbol{E}$ at all points in space, but clearly involves the derivatives of $\boldsymbol{E}$ rather than $\boldsymbol{E}$ itself. $\boldsymbol{E}$ is found by integration of eq. (1.37) *subject to the appropriate boundary conditions for the problem in hand.*

In view of our general comments regarding the usefulness of the scalar potential, it is often expedient to express eq. (1.37) in terms of V rather than $\boldsymbol{E}$. Since $\boldsymbol{E}=-\nabla V$, eq. (1.36) yields

$$-\nabla^2 V=\varrho/\epsilon_0$$

or

$$-\frac{\partial^2 V}{\partial x^2}-\frac{\partial^2 V}{\partial y^2}-\frac{\partial^2 V}{\partial z^2}=\varrho/\epsilon_0 \tag{1.38}$$

and eq. (1.38) is referred to as *Poisson's equation*. If the region under consideration contains no charge density, eq. (1.38) becomes

$$\nabla^2 V=0$$

$$\frac{\partial^2 V}{\partial x^2}+\frac{\partial^2 V}{\partial y^2}+\frac{\partial^2 V}{\partial z^2}=0 \tag{1.39}$$

and the general problem becomes one of integrating eqs (1.38) or (1.39) with the appropriate boundary conditions, Equation (1.39) is known as *Laplace's equation*, and such differential equations occur in many branches of science and engineering such as quantum mechanics, elasticity and aerodynamics. The methods for solving Laplace's equation form a subject by themselves and we do not consider them here.

Examples

1.1 Compare the gravitational and electrostatic forces between (a) two electrons and (b) between two uranium nuclei. What conclusion do you draw?

1.2 Show that

$$\nabla(\boldsymbol{p}\cdot\boldsymbol{r})=\boldsymbol{p}.$$

1.3 Show that

$$\nabla(1/r^3)=-3\boldsymbol{r}/r^5.$$

1.4 A thin circular ring of radius 0.1 m carries a uniformly distributed charge of 1 mC. What is the force on a point charge of magnitude 15 mC at the centre of the ring? What would be the force if the charge were moved 0.08 m along the ring axis?

1.5 Three charges Q_1, Q_2 and Q_3 are equally spaced along a line as shown

$$\underset{Q_1}{\cdot}\text{———————}\underset{Q_2}{\cdot}\text{———————}\underset{Q_3}{\cdot}$$

If the magnitude of Q_1 and Q_2 are equal, what must be the magnitude of Q_3 in order for the force on Q_1 to be zero?

1.6 Starting from the definition of electrostatic field, i.e. eq. (1.12), show that the field due to a thin uniform sheet of charge is given by $(\sigma/2\epsilon_0)\hat{\boldsymbol{n}}$, where σ is the uniform charge density and $\hat{\boldsymbol{n}}$ is a unit normal to the plane (cf. eq. (1.33)).

1.7 Starting from the equation for electrostatic potential (eq. (1.23)), find the electrostatic potential due to the sheet of charge in Example 1.6 above, and hence verify the expression for $\boldsymbol{E}$ using eq. (1.19).

1.8 A linear quadrupole consists of a charge $-2Q$ at the origin with charges $+Q$ at distances $\pm d$ along the z-axis. Show that, for points distant $r(r^3 \gg d^3)$ from the quadrupole centre

$$V = \frac{2Qd^2}{4\pi\epsilon_0 r^3}\left(\frac{3\cos^2\theta - 1}{2}\right),$$

where the angle θ is the angle between the point, the quadrupole centre and an end.

Note. The electrostatic potential due to a linear quadrupole thus varies as $1/r^3$ and the electric field turns out to vary as $1/r^4$. The fields of the three charges cancel almost exactly for $r \gg d$. The dipole and quadrupole potentials are two of a series of *multipole potentials*: for a general multipole, characterized by the letter l, the potential varies as $1/r^{l+1}$ and the electric field intensity as $1/r^{l+2}$. We will meet the multipole expansion in Chapter 8 where it is used as a basis for the expansion of an arbitrary potential.

1.9 For an array of point charges Q_i with position vectors $\boldsymbol{R}_i$, the electric dipole moment $\boldsymbol{p}_e$ is defined as

$$\boldsymbol{p}_e = \sum Q_i \boldsymbol{R}_i.$$

Show that the electric dipole moment is independent of the choice of coordinate origin only if $\sum Q_i = 0$.

1.10 Use Gauss' theorem to calculate the electric field at a point 100 pm from the nucleus of a ground-state hydrogen atom. Treat the nucleus as a point charge and the electronic wavefunction as

$$\sqrt{\left(\frac{1}{\pi a_0{}^3}\right)}\ \exp\left(-r/a_0\right) \qquad \text{where } a_0 = 52.9 \text{ pm}.$$

1.11 A charge of magnitude Q is distributed through the volume of a sphere of radius R with charge density

$$\varrho(r) = A(R - r) \qquad 0 \leqslant r \leqslant R.$$

Express A in terms of Q and R, and deduce expressions for the electric field at positions inside and outside the sphere.

1.12 A certain spherical charge distribution, which applies roughly to light nuclei, is given by

$$\varrho = \varrho_0(1 - r^2/a^2) \qquad r \leqslant a$$

$$\varrho = 0 \qquad r > a$$

Find the total charge as a function of a and ϱ_0 and use Gauss' thereom to calculate the electric field at points inside and outside the sphere.

1.13 If the electric field due to a certain atomic particle were proportional to $r^k\mathbf{r}$ $(k \neq -3)$, deduce whether the field would be a conservative field and deduce whether Gauss' theorem would apply.

1.14 Use Gauss' theorem to demonstrate that the electrostatic field at a distance r from a long straight wire carrying a charge γ per unit length is

$$\gamma/2\pi\epsilon_0 r$$

The measurement of electric quadrupole moments requires an electric field gradient, which may be produced by a four-wire capacitor, consisting of four long parallel wires, one through each corner of a square perpendicular to their length. The diagonal of the square is $2a$. The wires at opposite ends of one diagonal each carry charges per unit length γ, and the other wires each carry charges per unit length $-\gamma$. Calculate the field at a point on each diagonal a distance r from the intersection of the diagonals, assuming $r^2 \ll a^2$. Hence evaluate the field gradient along the diagonals at their intersection.

1.15 Assuming that the charge Q of a nucleus is distributed uniformly through a sphere of radius a, calculate the electric field at a distance r from the centre of the sphere at points both inside and outside the sphere. Hence show that the electrostatic potential inside the sphere is given by

$$V = (Q/8\pi\epsilon_0 a)\,(3 - (r/a)^2).$$

If the potential at the same distance from a point charge Q is V', show that the volume average of $(V - V')$ within the nucleus depends on the nuclear radius a. How does this account for isotope shifts in atomic spectra? (Recall that s electrons produce a non-zero charge density within the nucleus.)

Chapter 2

Electrostatics in the presence of materials

2.1 Introduction

In Chapter 1 we studied the laws of electrostatics in free space. We now give a 'non-molecular', or classical electromagnetic, account of the behaviour of materials in electrostatic fields and show how the laws need to be modified to take account of this behaviour. Throughout this text the terms *microscopic* (or *differential*) and *molecular* have rather different meanings. As noted earlier, classical electromagnetism is concerned with phenomena occurring on a length scale where the laws of quantum mechanics are 'unimportant': for example, in equations such as Gauss' theorem

$$\oint_S \boldsymbol{E}\cdot \mathrm{d}\boldsymbol{S} = \frac{1}{\epsilon_0}\int \varrho \, \mathrm{d}\tau \tag{2.1}$$

the surface and volume elements $\mathrm{d}S$ and $\mathrm{d}\tau$ should for the moment be interpreted as being either homogeneous molecular averages, or sufficiently large compared to molecular dimensions to contain very many molecules. In Parts B and C we show how the properties described here are related to molecular properties, and in Chapter 6 we examine the validity of the four laws of electromagnetism, Maxwell's equations, at the molecular level.

We differentiate between *conductors* and *insulators* (or *dielectrics*), a conductor being a material inside which electric charge can flow freely. We first summarize the properties of conductors and then in the remainder of the chapter outline the properties of dielectrics.

Inside a charged conductor the electric field is zero and all points are at the same potential when the charges present have attained their equilibrium positions, for if this were not true the charges would move in their own field until the resultant field became zero. Similar considerations for points on the surface of a charged conductor lead to the conclusions that

(1) The electric field at the surface of a charged conductor is normal to the surface

(2) The potential is constant at all points on the surface of a hollow conductor

Applying Gauss' theorem to an arbitrary surface completely contained within a hollow conductor leads to the conclusion that

(3) The net charge inside a hollow conductor is zero, so whenever net charge exists it must reside entirely on the surface of such a conductor.

2.2 Capacitance

A *capacitor* is a device for storing electric charge. A simple example is the *parallel plate* capacitor illustrated in Fig. 2.1 consisting of two conducting plates carrying equal and opposite charges $\pm q$. There is, therefore, no net charge on the capacitor, and if there is a potential difference $V = V_2 - V_1$ between the plates, the ratio q/V is defined to be the *capacitance C*. C is a purely geometric factor depending only on the physical arrangement of the plates, their sizes, etc. and *in principle* C is calculable from such knowledge, as the following simple example shows for the parallel plate capacitor.

In order to find C it is necessary to find V, which can in turn be found by calculating the field $\boldsymbol{E}$ between the plates. By a simple extension of example 1.6,

$$\boldsymbol{E} = \frac{\sigma}{\epsilon_0}\hat{\boldsymbol{x}} \tag{2.2}$$

and to find the change in potential we recall that $\boldsymbol{E} = -\nabla V$: hence

$$-E_x = \frac{\partial V}{\partial x}$$

and so

$$\begin{aligned} V(\text{positive plate}) - V(\text{negative plate}) &= E_x d \\ &= \sigma d/\epsilon_0. \end{aligned}$$

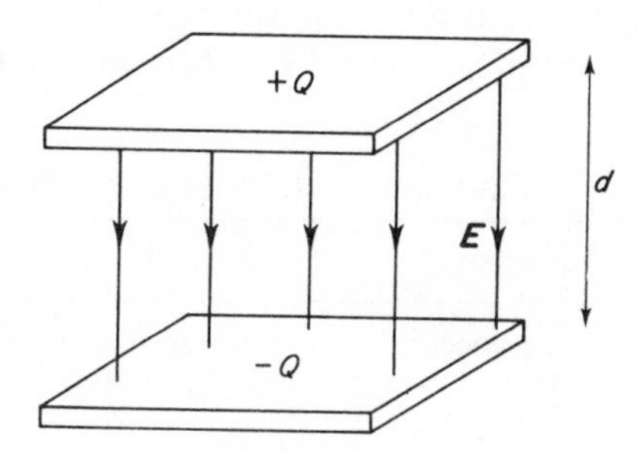

Fig. 2.1 A parallel-plate capacitor

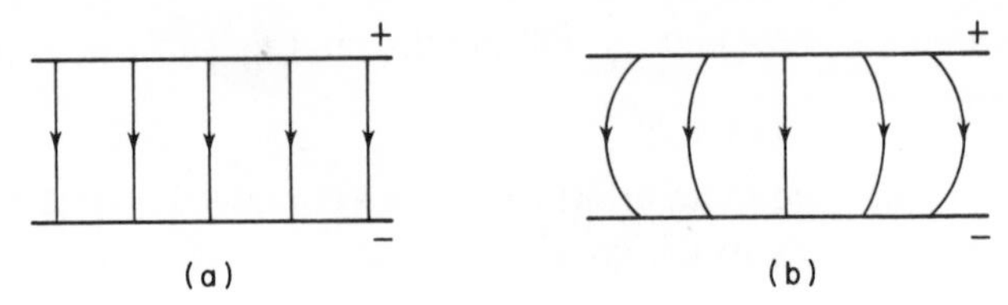

Fig. 2.2 Electrostatic flux lines for (a) an idealized parallel-plate capacitor and (b) a real device. Note the 'edge effect'

So from the definition of capacitance

$$C = A\epsilon_0/d \tag{2.3}$$

where A is the area of each plate. In practice there is both a physical limit on d because as $d \to 0$ electrical breakdown of the material between the plates occurs, and an end-effect due to the plates being of non-infinite dimensions. This end-effect is shown in Fig. 2.2, together with the idealized field of eq. (2.2).

2.3 Polarization

In 1837 Faraday showed that when the space between the plates of a capacitor was filled with a dielectric such as glass or mica, its capacitance became greater. The multiplicative factor ϵ_r is called the *relative permittivity* of the material, and naturally materials are classified as dielectrics according to their ϵ_r. Representative values of ϵ_r for some dielectrics are shown in Table 2.1. ϵ_r is an experimentally determined quantity, its older name being the *dielectric constant K*. Relative permittivities are generally different when measured with static and with alternating voltages applied to the plates, and are also dependent on the frequency of such alternating voltages. The dependence of ϵ_r on the frequency of an electric field becomes very important when considering the relatively high frequencies associated with optical electromagnetic waves.

Dielectric materials such as glass and mica differ from metallic conductors in that they have no free electrons which can move through the material under the influence of an external electric field. When an electrostatic field $\boldsymbol{E}(\boldsymbol{r})$ is

Table 2.1 Static relative permittivities at room temperature

Substance	ϵ_r
'Free space'	1
Air	1.0006
Glass	~6
Water	81

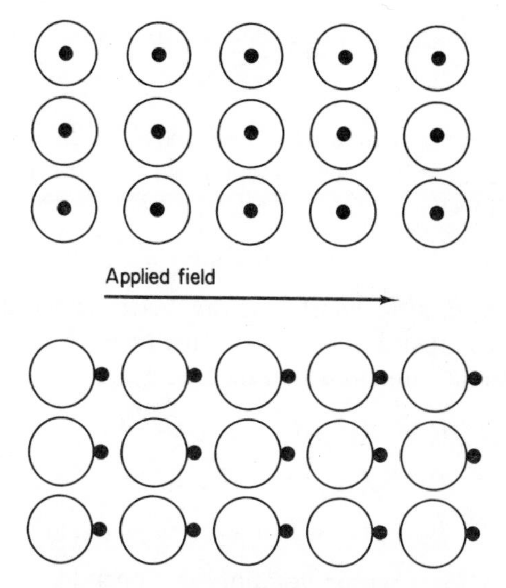

Fig. 2.3 The effect of an applied electrostatic field on a 'crystal' composed of atoms. The full circles represent nuclei and the outer circles electron density. Polarization gives rise to a separation of charge and hence a dipole moment is induced in the sample

applied to a slab of dielectric as illustrated in Fig. 2.3, the bound electrons are displaced slightly in the direction away from the field whilst the nuclei are displaced a correspondingly smaller distance into the direction of $\boldsymbol{E}$. The dielectric is said to be *polarized* by the field and as a result of this polarization the molecules acquire an induced dipole moment. If the sample is polarized parallel to $\boldsymbol{E}$ simple considerations show that the bulk of the dielectric remain electrically neutral whilst the left-hand face of the slab illustrated in Fig. 2.3 will acquire a net excess negative charge equal and opposite to that acquired by the right-hand face.

For the remainder of this chapter we discuss polarization from the standpoint of classical electromagnetism, returning to the molecular description in Chapter 8. Thus, attention focuses on the element of induced dipole $\mathrm{d}\boldsymbol{p}$ where

$$\mathrm{d}\boldsymbol{p} = \boldsymbol{P}\,\mathrm{d}\tau \tag{2.4}$$

with volume element sufficiently large on the molecular scale as to be homogeneous. It seems reasonable to assume that a functional relationship $\boldsymbol{P} = \boldsymbol{P}(\boldsymbol{E})$ will exist between the polarization $\boldsymbol{P}$ and field $\boldsymbol{E}$, and this is the case. Dielectrics are classified according to their response to the external field, but of course this response must be measured experimentally. In the simplest possible case, the polarization may be directly proportional to $\boldsymbol{E}$

$$P \propto E \tag{2.5}$$

in which case one speaks of a *linear* dielectric. If in addition the electrical properties of the dielectric are independent of direction the dielectric is referred to as a linear *isotropic* dielectric, and

$$\boldsymbol{P} = \chi_e \epsilon_0 \boldsymbol{E} \tag{2.6}$$

where χ_e is the *electric susceptibility* of the dielectric and is a scalar quantity. Except for materials that are either isotropic or have cubic symmetry, $\boldsymbol{P}$ and $\boldsymbol{E}$ may have different directions in which case

$$\begin{aligned} P_x &= \epsilon_0(\chi_{e,\,xx}E_x + \chi_{e,\,xy}E_y + \chi_{e,\,xz}E_z) \\ P_y &= \epsilon_0(\chi_{e,\,yx}E_x + \chi_{e,\,yy}E_y + \chi_{e,\,yz}E_z) \\ P_z &= \epsilon_0(\chi_{e,\,zx}E_x + \chi_{e,\,yz}E_y + \chi_{e,\,zz}E_z) \end{aligned} \tag{2.7}$$

or in the more compact tensor notation of Appendix A.14

$$\boldsymbol{P} = \epsilon_0 \mathbf{X}_e \cdot \boldsymbol{E}, \tag{2.8}$$

Again, *non-linear* dielectrics exist for which

$$\boldsymbol{P} = \epsilon_0 \mathbf{X}_e \cdot \boldsymbol{E} + \text{terms in } E^2 + \ldots \tag{2.9}$$

and we stress that the correct classification for a given dielectric can only be determined by experimental measurements. All dielectrics, however, become linear as $\boldsymbol{E} \to \boldsymbol{0}$. A final possibility is that the permanent molecular electric dipole moments of some molecules are partially aligned even in the absence of an external field. Such materials, which are the electrostatic analogues of permanent magnets, are called *electrets*.

2.4 Polarization charges

The effect of an external electric field on a dielectric is to induce a dipole moment $\boldsymbol{P}\,\mathrm{d}\tau$ in each volume element $\mathrm{d}\tau$. We now show that, as far as points *outside* the dielectric are concerned, the resulting potential is exactly what would be obtained from a certain volume distribution of charge throughout the dielectric, and a certain distribution of charge on the surface of the sample.

Thus, we require to calculate the potential at field point $\boldsymbol{R}$ of Fig. 2.4. From eq. (1.26) we know that the potential at $\boldsymbol{R}$ due to the dipole $\mathrm{d}\boldsymbol{p} = \boldsymbol{P}\,\mathrm{d}\tau$ is

$$\mathrm{d}V = \frac{\mathrm{d}\boldsymbol{p}\cdot(\boldsymbol{R}-\boldsymbol{r})}{4\pi\epsilon_0\,|\boldsymbol{R}-\boldsymbol{r}|^3}, \tag{2.10}$$

where the field point $\boldsymbol{R}$ is regarded as a fixed point and the coordinates $\boldsymbol{r}$ of the point inside the dielectric are regarded as variables, because it will be

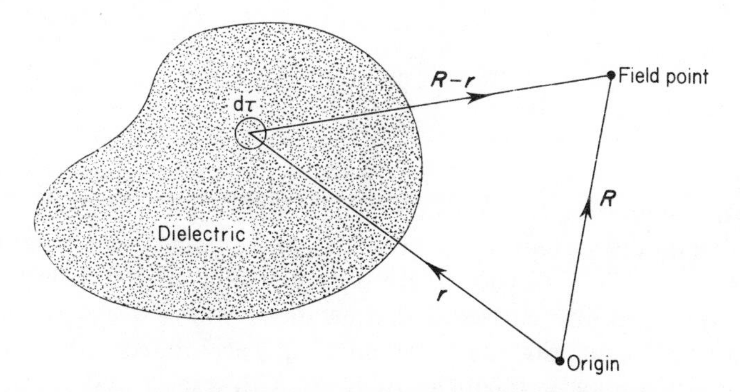

Fig. 2.4 Calculation of the potential at field point R due to a polarized sample of dielectric. The polarization can be thought of in terms of the volume and surface polarization charges

necessary to integrate eq. (2.10) to find the full potential at $\boldsymbol{R}$. Putting

$$\boldsymbol{\nabla} = \frac{\partial}{\partial x}\boldsymbol{i} + \frac{\partial}{\partial y}\boldsymbol{j} + \frac{\partial}{\partial z}\boldsymbol{k}$$

it is easily verified (Example 2.1) that

$$\boldsymbol{\nabla}\left|\frac{1}{\boldsymbol{R}-\boldsymbol{r}}\right| = \frac{\boldsymbol{R}-\boldsymbol{r}}{|\boldsymbol{R}-\boldsymbol{r}|^3} \tag{2.11}$$

so that

$$4\pi\epsilon_0 \mathrm{d}V = \boldsymbol{P}\cdot\boldsymbol{\nabla}\left|\frac{1}{\boldsymbol{R}-\boldsymbol{r}}\right|\mathrm{d}\tau$$

and the potential at field point R is given by eq. (2.12):

$$4\pi\epsilon_0 V = \int \boldsymbol{P}\cdot\boldsymbol{\nabla}\left|\frac{1}{\boldsymbol{R}-\boldsymbol{r}}\right|\mathrm{d}\tau. \tag{2.12}$$

Using the vector identity

$$\boldsymbol{\nabla}\cdot\frac{\boldsymbol{P}}{|\boldsymbol{R}-\boldsymbol{r}|} = \boldsymbol{P}\cdot\boldsymbol{\nabla}\left|\frac{1}{\boldsymbol{R}-\boldsymbol{r}}\right| + \left|\frac{1}{\boldsymbol{R}-\boldsymbol{r}}\right|\boldsymbol{\nabla}\cdot\boldsymbol{P}$$

eq. (2.12) can be rewritten

$$4\pi\epsilon_0 V = \left\{-\int\frac{\boldsymbol{\nabla}\cdot\boldsymbol{P}}{|\boldsymbol{R}-\boldsymbol{r}|}\mathrm{d}\tau + \int\boldsymbol{\nabla}\cdot\frac{\boldsymbol{P}}{|\boldsymbol{R}-\boldsymbol{r}|}\mathrm{d}\tau\right\} \tag{2.13}$$

and using the divergence theorem of Appendix A to replace the second volume integral on the right-hand side of eq. (2.13) by a surface integral we find

$$4\pi\epsilon_0 V = -\int\frac{\boldsymbol{\nabla}\cdot\boldsymbol{P}}{|\boldsymbol{R}-\boldsymbol{r}|}\mathrm{d}\tau + \int\frac{\boldsymbol{P}\cdot\mathrm{d}\boldsymbol{S}}{|\boldsymbol{R}-\boldsymbol{r}|} \tag{2.14}$$

or

$$4\pi\epsilon_0 V = \int \frac{\varrho_p \, d\tau}{|R - r|} + \int \frac{\sigma_p \, dS}{|R - r|}, \tag{2.15}$$

where $\varrho_p = -\nabla \cdot P$ and $\sigma_p = P \cdot n$. Thus, the potential due to a polarized sample *can be interpreted as* arising from a *surface polarization charge* σ_p of density $P \cdot n$ on the surface of the insulator, and a *volume* polarization charge ϱ_p of density $-\nabla \cdot P$ distributed through the bulk of the dielectric. If the dielectric is uniformly polarized, P is constant so $\nabla \cdot P = 0 = \varrho_p$. In calculating ϱ_p, it must be stressed that the differentiation is performed with respect to the coordinates of a point in the dielectric r, not with respect to the coordinate of the field point R. The charge densities arise at the molecular level from the bound electrons and protons in the dielectric and for that reason are often referred to as the *bound* charge densities σ_b and ϱ_b.

2.5 The electric displacement, *D*

When a sample of dielectric is inserted between the plates of a capacitor, the sign of the induced charges is *always* such as to reduce the field inside the dielectric compared to the vacuum field. Figure. 2.5 shows the effect of inserting a dielectric of relative permittivity $\epsilon_r = 2$, hence reducing the field inside the dielectric by a factor of 2, as we show formally in Section 2.6. When using Guass' theorem, for example, to investigate electric fields in the presence of dielectrics it is necessary to include contributions from the polarization charges in the right-hand side of eq. (2.16), where $Q_{total} = Q_{free} + Q_{polarization}$, the first term often being called the *external* charge. Now eq. (1.31) can be written

$$\oint E \cdot dS = Q_{total}/\epsilon_0 \tag{2.16}$$

or

$$\int (\epsilon_0 E + P) \cdot dS = Q_{free} \tag{2.17}$$

and since the right-hand side of eq. (2.17) contains only the free charges it is

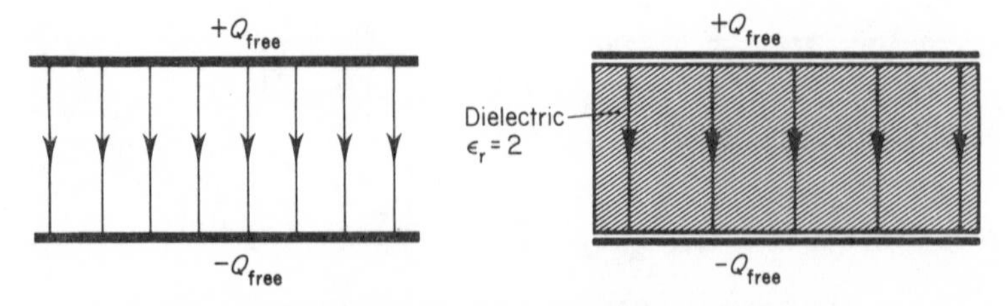

Fig. 2.5 The effect of a dielectric of relative permittivity 2 on the electrostatic field inside a capacitor. The electric flux is reduced by a factor 2 inside the dielectric

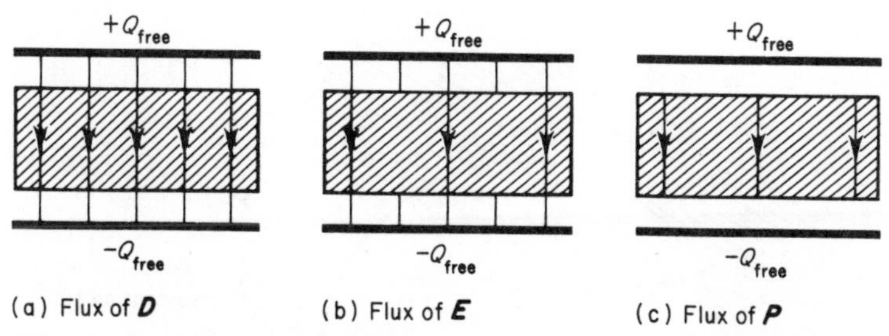

Fig. 2.6 Illustrating the differences between **D** and **P**. **P** only exists inside the dielectric and **D** is continuous across boundaries

both traditional and convenient to regard $\epsilon_0 \boldsymbol{E} + \boldsymbol{P}$ as a more useful quantity than $\boldsymbol{E}$ because the polarization charges are inherent in the nature of the dielectric whereas the distribution of the free charges can be externally controlled. The vector field

$$\boldsymbol{D} = \epsilon_0 \boldsymbol{E} + \boldsymbol{P} \tag{2.18}$$

is called the *electric displacement*, and Gauss' theorem can therefore be rewritten

$$\oint \boldsymbol{D} \cdot \mathrm{d}\boldsymbol{S} = Q_{\text{free}} \tag{2.19}$$

or, alternatively in its differential form

$$\nabla \cdot \boldsymbol{D} = \varrho_{\text{free}}. \tag{2.20}$$

It is, however, important to distinguish between $\boldsymbol{D}$ and $\boldsymbol{E}$: the electric field $\boldsymbol{E}$ has a clear physical meaning irrespective of the presence of materials. The electric displacement is introduced in order to work with a vector field whose divergence depends only on the *free* charge density. The dimensions of $\boldsymbol{D}$ are those of $\epsilon_0 \boldsymbol{E}$ or $\boldsymbol{P}$, $\mathrm{C\,m^{-2}}$.

Figure 2.6 summarizes pictorially the relationship between $\boldsymbol{E}$, $\boldsymbol{P}$ and $\boldsymbol{D}$: $\boldsymbol{D}$ is a constant field, $\boldsymbol{E}$ is reduced by a factor of ϵ_r within the dielectric and $\boldsymbol{P}$ exists only within the sample of dielectric.

2.6 Some properties of D

(1) In linear isotropic media where $\boldsymbol{P} = \epsilon_0 \chi_e \boldsymbol{E}$

$$\boldsymbol{D} = \epsilon_0 \boldsymbol{E} + \boldsymbol{P} = \epsilon_0 (1 + \chi_e) \boldsymbol{E}.$$

Hence

$$\boldsymbol{D} = \epsilon \boldsymbol{E}, \tag{2.21}$$

where $\epsilon = \epsilon_0 \epsilon_r$ is the *permittivity* of the medium, and $\epsilon_r = (1 + \chi_e)$.

(2) Suppose that a single point charge Q is located in a linear isotropic medium of permittivity ϵ.

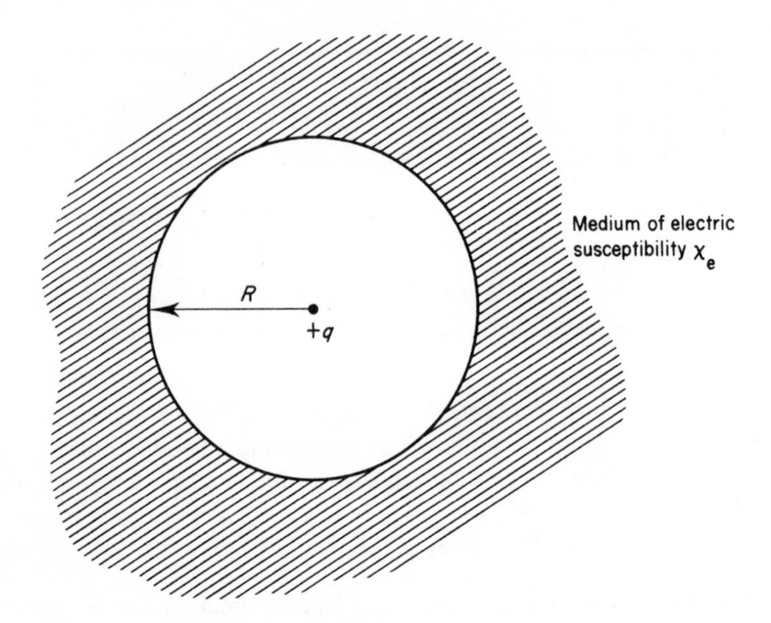

Fig. 2.7 Construct for the calculation of **D** inside a dielectric

Applying the appropriate form of Gauss' theorem to a sphere of radius R surrounding Q,

$$4\pi R^2 D = Q,$$

so the electric field is

$$E = Q/4\pi R^2 \epsilon$$

and it follows that the force between charges q_1 and q_2 in a medium of permittivity ϵ is

$$F = \frac{Q_1 Q_2}{4\pi\epsilon} \frac{(r_1 - r_2)}{|r_1 - r_2|^3} \tag{2.22}$$

or, the force is reduced by a factor of ϵ_r from its vacuum level.

(3) Returning to the parallel plate capacitor of Section 2.2 it is easily seen that the capacitance is given by

$$C = \frac{\epsilon_0(1 + \chi_e)}{d} A = \frac{\epsilon A}{d}$$

when a linear isotropic dielectric is present. That is, the capacitance increases by a factor $(1 + \chi_e)$, and hence

$$(1 + \chi_e) = K = \epsilon_r$$

showing the relationship between the dielectric constant, relative permittivity and electric susceptibility.

2.7 Electrostatic energy

In this final section we show that the energy density stored in an electrostatic charge distribution is $\frac{1}{2}\boldsymbol{D}\cdot\boldsymbol{E}$. The following simple argument indicates the truth of the assertion: consider the work done in charging up a capacitor of capacity C. Suppose that at a given instant the potential difference between the plates is V' and that chage $\mathrm{d}q'$ is transferred from one plate to the other under the constant potential V'. The work done is

$$\mathrm{d}w = V'\mathrm{d}q' = q'\mathrm{d}q'/C \tag{2.23}$$

and by integration of eq. (2.23) the *total* work done when the plates finally carry charges $\pm\, q$ is

$$w = \tfrac{1}{2}q^2/C = \tfrac{1}{2}CV^2. \tag{2.24}$$

The above discussion is generally true for any capacitor. In the special case of a parallel plate capacitor where $E = \sigma/\epsilon$ and $C = \epsilon A/d$, eq. (2.24) becomes

$$w = \tfrac{1}{2}\epsilon E^2 Ad. \tag{2.25}$$

The work done per unit volume, or *energy density*, is thus $\frac{1}{2}\epsilon E^2$ or $\frac{1}{2}\boldsymbol{D}\cdot\boldsymbol{E}$, and we now show that this is a general result for the energy density associated with any electrostatic field.

Suppose point charges q_1 and q_2 are at position vectors $\boldsymbol{r}_1$ and $\boldsymbol{r}_2$. The work done in bringing q_1 from infinity to $\boldsymbol{r}_1$ in the field of q_2 is

$$U_{1,2} = \frac{q_1 q_2}{4\pi\epsilon_0 r_{12}} \tag{2.26}$$

which by symmetry is exactly the work done in bringing q_2 from infinity to $\boldsymbol{r}_2$ in the field of q_1. We thus speak of the *mutual potential energy* $U_{1,2}$ of the system of (two) charges. For an array of point charges $q_1, q_2, \ldots, q_n$ the mutual potential energy is

$$U = \sum_{i>j}\sum U_{i,j} \tag{2.27}$$

where

$$U_{i,j} = \frac{q_i q_j}{4\pi\epsilon_0 r_{ij}}$$

and the summation in eq. (2.27) only counts pairs of interactions once. Equation (2.27) can be written

$$U = \frac{1}{2}\sum_{i\neq j}\sum U_{i,j} \tag{2.28}$$

and from the definition of the potential due to a point charge we see that

$$U = \frac{1}{2}\sum q_i V(r_i) \tag{2.29}$$

where $V(\boldsymbol{r}_i)$ is the potential at $\boldsymbol{r}_i$ due to all the other charges $q_j (j \neq i)$ In the case of a continuous charge distribution, eq. (2.29) becomes

$$U = \frac{1}{2}\int \varrho(\boldsymbol{r}) V(\boldsymbol{r})\, \mathrm{d}\tau, \tag{2.30}$$

where for the moment we have assumed that no dielectric material is present. Thus Gauss' theorem

$$\nabla \cdot \boldsymbol{E} = \varrho(\boldsymbol{r})/\epsilon_0$$

gives

$$U = \frac{\epsilon_0}{2}\int (\nabla \cdot \boldsymbol{E})\, V(\boldsymbol{r})\, \mathrm{d}\tau \tag{2.31}$$

and since

$$\nabla \cdot (V\boldsymbol{E}) = V(\nabla \cdot \boldsymbol{E}) + \boldsymbol{E} \cdot \nabla V$$

eq. (2.31) can be rewritten

$$U = \frac{\epsilon_0}{2}\int \{\nabla \cdot (V\boldsymbol{E}) - \boldsymbol{E} \cdot \nabla V\}\, \mathrm{d}\tau$$

or

$$U = \frac{\epsilon_0}{2}\int \boldsymbol{E} \cdot \boldsymbol{E}\, \mathrm{d}\tau + \frac{\epsilon_0}{2}\int \nabla \cdot (V\boldsymbol{E})\, \mathrm{d}\tau$$

since $\boldsymbol{E} = -\nabla V$ by definition.

The second term on the right-hand side can be rewritten

$$\frac{\epsilon_0}{2}\oint_S V\boldsymbol{E} \cdot \mathrm{d}\boldsymbol{S}$$

using the divergence theorem of Appendix A, where the surface S can be chosen to be an essentially infinite sphere. This term is zero so that

$$U = \frac{\epsilon_0}{2}\int \boldsymbol{E} \cdot \boldsymbol{E}\, \mathrm{d}\tau. \tag{2.32}$$

Hence the energy density is $\frac{1}{2}\epsilon_0 E^2$ or $\frac{1}{2}\boldsymbol{D} \cdot \boldsymbol{E}$ for free space. If dielectric material *is* present, eq. (2.30) becomes

$$U = \frac{1}{2}\int \varrho_{\mathrm{f}}(\boldsymbol{r}) V(\boldsymbol{r})\, \mathrm{d}\tau,$$

where $V(\boldsymbol{r})$ contains contributions from the polarization charges in addition to the free charges. Using $\varrho_{\mathrm{f}} = \nabla \cdot \boldsymbol{D}$ and following the arguments given above yields an energy density $\frac{1}{2}\boldsymbol{D} \cdot \boldsymbol{E}$.

Examples

2.1 Verify that

$$\nabla \left| \frac{1}{R-r} \right| = \frac{R-r}{|R-r|^3},$$

where ∇ operates only on the coordinates of r. What is the corresponding result when ∇ operates only on the coordinates of R?

2.2 Use Gauss' theorem to find the electric field at points outside a spherical charged hollow conductor. Hence show that the capacitance of a pair of such concentric spheres is

$$4\pi\epsilon_0 \frac{ab}{b-a},$$

where the radii a, b are such that $b > a$.

2.3 Show that the capacitance of two coaxial conducting cylinders of radii a, b and length L is

$$\frac{2\pi\epsilon_0 L}{\log_e (b/a)} \qquad b > a.$$

2.4 Two molecules whose electric dipole strengths are p, p' are fixed at points A, B at a distance d apart with both dipoles pointing in the same direction AB. Show that there is an attractive force between them of

$$6pp'/4\pi\epsilon_0 d^4.$$

(*Hint*. Take the dipoles as charges $-q$, $+q$ at $x = -l/2$, $x = l/2$ and $-q$, q' at $x = d - l'/2$, $x = d + l'/2$, where $lq = p$, $l'q' = p'$ and let l and l' become very small.)
If two water molecules are oriented in this way, calculate the dipole–dipole interaction when their centres are at a separation of 300 pm. ($p_e = 5.97 \times 10^{-30}$ C m).

2.5 Consider a block of dielectric with bound charge densities ϱ_b, σ_b. Show that

$$\int_V \varrho_b \, d\tau + \oint_S \sigma_b \, dS = 0,$$

where V is the volume of the dielectric and S is the surface. This demonstrates that the total net bound charge is zero.

2.6 Using the result of Example 2.2 and by letting the radius of the outer sphere tend to infinity, show that the capacitance of an isolated charged sphere of radius r is $4\pi\epsilon_0 r$. In the Wilson cloud chamber, droplets of liquid condense preferentially around charged particles, so making the position of the particles visible. The total energy of a spherical droplet of radius r carrying a charge q is given by the sum of its electrostatic energy and its surface energy $4\pi r^2\gamma$, where γ is the surface tension of the liquid. Show that for sufficiently small radii the total energy decreases with increasing radius because of the electrostatic term, and hence show that the equilibrium volume of the droplet is $q^2/48\ \pi\epsilon_0\gamma$ (electrostatic energy $= \frac{1}{2}CV^2$).

2.7 Show that the capacitance per unit length of the coaxial cable shown below is

$$\frac{2\pi\epsilon_0}{\dfrac{1}{\epsilon_1}\log_e\dfrac{r_2}{r_1}+\dfrac{1}{\epsilon_2}\log_e\dfrac{r_3}{r_2}}$$

where ϵ_1 and ϵ_2 are the relative permittivities of the intervening dielectrics.

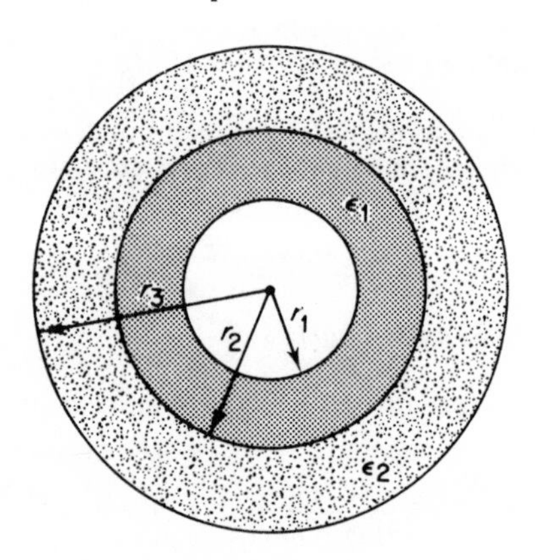

Chapter 3

Magnetostatics

3.1 Introduction

So far our concern has been with electric charges at rest. If electric charges move so that there is a net drift of charge across a surface we become concerned with *conduction currents* (named so as to distinguish them from the *displacement currents* to be introduced in Chapter 7). Conduction currents are very often simply referred to as electric currents.

In this chapter our concern is with the magnetic effects of *steady* currents and in Chapter 5 we examine the fields associated with electric currents that vary with time. For simplicity we start our discussion of magnetostatics by discussing the magnetic effects of electric currents in free space: that is, we assume that no magnetic materials are present between the electric circuits carrying the currents. In Chapter 4 we give the classical electromagnetic description of the properties of magnetic materials, returning to the molecular interpretation in Chapter 10.

3.2 Ampère's Law

Around 1820, Oersted discovered experimentally that electric currents could exert forces similar to those exerted by permanent magnets, for example on a compass needle. Ampère then showed that an electric circuit could exert just such a force on another electric circuit. The mutual force between the two electrical circuits cannot be the same force as the electrostatic force discussed in Chapter 1, because each circuit contains an equal amount of positive and negative charge. It is our belief that the magnetic effects observed by Oersted are due entirely to the movement of electric charges in the electric circuits.

In electrostatics it was easy to realize that the fundamental particles responsible for electric charge are electrons and protons, and it is natural to ask whether any fundamental particles are responsible for magnetism. Such particles will be referred to as *magnetic monopoles*.

In 1269 Peregrinus noted that lines of force around a lodestone were concentrated at two points which he designated the 'north' and the 'south' poles of the magnet. Subsequent observation has confirmed that all magnetic objects have regions of opposite polarity. The field of magnetic monopoles has achieved great importance in recent years; magnetic monopoles were postulated by P. A. M. Dirac in 1931 in order to explain the quantization of electric charge. He established the basic relation

$$g = \frac{hc}{4\pi e} n$$

between the elementary charge e and the hypothesized magnetic charge g, where n is an integer which in the original proposal could take values $n = \pm 1, \pm 2, \ldots$. Such monopoles are referred to as 'classical' monopoles and fruitless searches have been made for these monopoles at many high-energy accelerators.

In 1974 't Hooft and Polyakov showed that magnetic monopoles can theoretically exist as stable solutions in some special theories, and this discovery has caused a resurgence of interest in 'monopole hunting'. No positive experimental findings have been reported to date, and it is thus our belief that *all* magnetic phenomena are caused by charged particles. If magnetic monopoles are ever detected experimentally, it will be necessary to modify the basic equations of electromagnetism to describe phenomena involving these particles (see Example 6.2, page 85).

The question as to what molecular feature is responsible for magnetism in permanent magnets will be dealt with in Chapter 10: we merely remark here that the existence of angular momentum $\boldsymbol{M}$ causes a charged particle to have an associated magnetic dipole moment $\boldsymbol{p}_m \propto \boldsymbol{M}$. Such angular momentum cannot be properly described by classical mechanics or electromagnetism, and is of two kinds: orbital angular momentum and spin angular momentum. Spin angular momentum can only be properly explained by Dirac's relativistic quantum mechanics. If the molecular magnetic dipoles are naturally partially aligned, the sample of material will have an overall magnetic dipole moment, like the permanent electric dipole possessed by the electrets discussed in Chapter 2.

Although the magnetostatic force law is of necessity one involving the interaction of complete circuits, it often proves useful to express the law in terms of differential portions of each complete circuit. Thus if C_a and C_b are complete circuits $I_a \mathrm{d}\boldsymbol{l}_a$ is a *current element* of circuit a, where I_a is the current flowing through C_a. In Fig. 3.1 the batteries producing the currents are shown some distance away from the parts of the circuits under study. One of Ampère's first experiments showed that two oppositely directed currents very close together spatially produced no net effect on a third circuit. Ampère's experiments showed that the forces between C_a and C_b depended on:

(1) I_a, I_b and their direction of flow. For an idealized circuit of two long parallel wires, the force is repulsive if I_a and I_b flow in the same direction and attractive if they flow in opposite directions.
(2) The lengths and relative orientations and geometries of C_a and C_b.

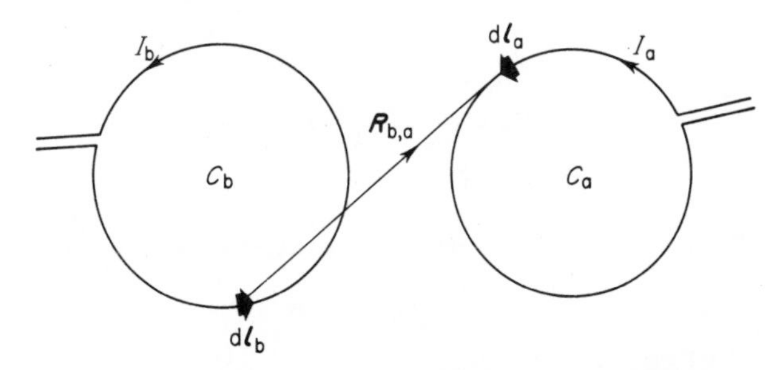

Fig. 3.1 Geometric construct used in Ampère's law for the mutual force between the circuits C_a and C_b. It was noted experimentally at a very early stage that the magnetic effects of two neighbouring wires carrying equal and opposite currents cancelled each other. The sources of the currents in either circuit are at the ends of the long neighbouring wires

(3) The angle between current elements.
(4) R^{-2}, where R is the distance between current elements.

The basic experimental law for the force $F_{b,a}$ on circuit C_a due to C_b can be written as

$$\boldsymbol{F}_{b,a} = \frac{\mu_0}{4\pi} \oint_{C_a} \oint_{C_b} \frac{I_a \, d\boldsymbol{l}_a \times (I_b \, d\boldsymbol{l}_b \times \boldsymbol{R}_{b,a})}{R_{b,a}^3} \tag{3.1}$$

and this expression is known as *Ampère's law*. Of course Ampère's simple measurements could not possibly have led to such a complicated expression: eq. (3.1) is a generalization of Ampère's results, and the force is thus seen to be a double line integral over the two, *complete*, interacting circuits. In SI the proportionality constant $\mu_0/4\pi$ is *defined* to be 10^{-7} N A^{-2} (= H m^{-1}), where μ_0 is the *permeability of free space*. Since the newton and the metre are fundamental SI constants this choice of μ_0 fixes the definition of the coulomb C and hence the ampere A = C s^{-1}.

If several circuits $C_b, C_c, \ldots, C_n$ can interact with C_a then the superposition principle applies as in electrostatics

$$\boldsymbol{F}_a = \sum_{k=1}^{n} \boldsymbol{F}_{k,a}. \tag{3.2}$$

At first sight eq. (3.1) appears not to conform to Newton's third law on account of the lack of symmetry in the integrand. Equation (3.1) can, however, be written in a more symmetrical form in order to demonstrate explicitly that $\boldsymbol{F}_{b,a} = -\boldsymbol{F}_{a,b}$. Expanding the vector triple product gives

$$d\boldsymbol{l}_a \times (d\boldsymbol{l}_b \times \boldsymbol{R}) = (d\boldsymbol{l}_a \cdot \boldsymbol{R}) \, d\boldsymbol{l}_b - \boldsymbol{R} \, (d\boldsymbol{l}_a \cdot d\boldsymbol{l}_b)$$

where $\boldsymbol{R} = \boldsymbol{r}_a - \boldsymbol{r}_b$. Putting

$$\nabla_a = \frac{\partial}{\partial x_a}\boldsymbol{i} + \frac{\partial}{\partial y_a}\boldsymbol{j} + \frac{\partial}{\partial z_a}\boldsymbol{k}$$

from which

$$\nabla_a\left(\frac{1}{R}\right) = -\frac{\boldsymbol{R}}{R^3}$$

the integrand of eq. (3.1) becomes

$$\frac{d\boldsymbol{l}_a \times (d\boldsymbol{l}_b \times \boldsymbol{R})}{R^3} = -d\boldsymbol{l}_a \cdot \nabla_a\left(\frac{1}{R}\right) d\boldsymbol{l}_b - \frac{d\boldsymbol{l}_a \cdot d\boldsymbol{l}_b}{R^3}\boldsymbol{R}. \tag{3.3}$$

On performing the integration, the first term

$$-I_a I_b \oint_{C_b}\left\{\oint_{C_a} d\boldsymbol{l}_a \cdot \nabla_a\left(\frac{1}{R}\right)\right\} d\boldsymbol{l}_b \tag{3.4}$$

vanishes since the integral in parentheses is zero, being the integral over a *closed path* of the differential of a scalar. Hence

$$\boldsymbol{F}_{b,a} = -\frac{\mu_0}{4\pi} I_a I_b \oint_{C_a}\oint_{C_b} \frac{d\boldsymbol{l}_a \cdot d\boldsymbol{l}_b}{R^3}\boldsymbol{R} \tag{3.5}$$

which does now show explicitly that Newton's third law is satisfied: $\boldsymbol{F}_{b,a} = -\boldsymbol{F}_{a,b}$ since $\boldsymbol{R} = \boldsymbol{r}_a - \boldsymbol{r}_b$ changes sign on interchanging the labels 'a' and 'b'. Equation (3.5) is known as Neumann's formula.

However, despite the symmetry of eq. (3.5) it is more convenient to start the discussion of magnetostatic fields from eq. (3.1), which is more easily cast into a circuit C_a interacting with a certain field produced by C_b.

3.3 The magnetic induction

As in the electrostatic case, it proves useful to recast eq. (3.1) in terms of a vector field. The current elements $I_a\, d\boldsymbol{l}_a$ and $I_b\, d\boldsymbol{l}_b$ corresponding to C_a and C_b are shown in Fig. 3.1, and the integrand in eq. (3.1)

$$\frac{\mu_0}{4\pi} I_a\, d\boldsymbol{l}_a \times (I_b\, d\boldsymbol{l}_b \times \boldsymbol{R})/R^3$$

is interpreted as a force between the two current elements, the total force being the sum of all such elemental forces. The force on current element $I_a\, d\boldsymbol{l}_a$ due to the complete circuit C_b is therefore

$$I_a\, d\boldsymbol{l}_a \times \left[\frac{\mu_0}{4\pi} I_b \oint_{C_b} \frac{d\boldsymbol{l}_b \times \boldsymbol{R}}{R^3}\right] \tag{3.6}$$

which gives a field interpretation to the basic force law: the vector field at point R due to the circuit C_b is thus

$$\boldsymbol{B}_b = \frac{\mu_0}{4\pi} I_b \oint_{C_b} \frac{d\boldsymbol{l}_b \times \boldsymbol{R}}{R^3}, \tag{3.7}$$

where $\boldsymbol{B}_b$ is called the *magnetic induction*, and the direction of $\boldsymbol{R}$ is from the circuit C_b to the field point under study. Where no confusion can exist, the suffix b can be omitted and eq. (3.7) is referred to as the *Biot–Savart law*. The force on current element $I_a\, d\boldsymbol{l}_a$ is thus $I_a\, d\boldsymbol{l}_a \times \boldsymbol{B}_b$. The magnetic induction is the fundamental magnetostatic field, and is the analogue of $\boldsymbol{E}$ in electrostatics. In SI the units of $\boldsymbol{B}$ are tesla T, where $T = N/A\ m = Wb\ m^{-2}$. The older Gaussian system of units uses gauss for $\boldsymbol{B}$, where $1\ T = 10^4\ G$.

Elementary applications

We now illustrate how the Biot–Savart law, eq. (3.7), can be used directly in cases of high symmetry to find analytical expressions for the magnetic induction $\boldsymbol{B}$. As in Chapter 1, the examples are somewhat artificial and it rarely proves possible to find $\boldsymbol{B}$ by the direct application of eq. (3.7).

Magnetic induction due to long wire carrying steady current If the wire is infinite, α ranges from $-\pi/2$ to $\pi/2$. Unlike the electrostatic field example of Section 1.6, each current element makes a parallel contribution to $\boldsymbol{B}$ and there is no cancellation. Lines of constant $\boldsymbol{B}$ form concentric rings around the wire, and hence

$$\boldsymbol{B} = \frac{\mu_0 I}{2\pi a} \hat{\boldsymbol{n}} \tag{3.8}$$

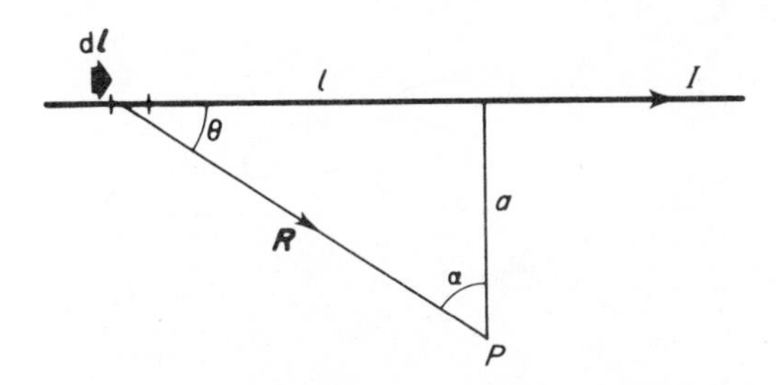

Fig. 3.2 Calculation of the magnetic induction due to a long wire carrying steady current I. $I\,d\mathbf{l}$ is a current element and the vector $\mathbf{R}$ points from the current element to the field point, which is taken as the centre of the coordinate system

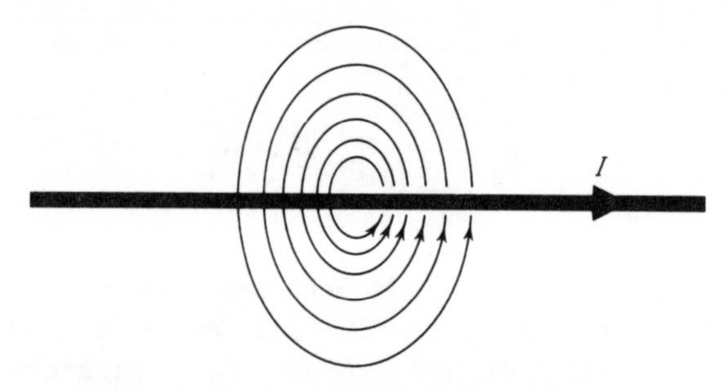

Fig. 3.3 Lines of B for the long wire carrying a steady current. Note that the flux lines do not have 'beginnings' or 'ends', in contrast to the electrostatic case

and the magnitude of $\boldsymbol{B}$ falls off as the first power of the distance from the wire. For a finite length wire the eq. (3.8) becomes

$$\boldsymbol{B}=\frac{\mu_0 I}{4\pi a}(\sin\alpha_1+\sin\alpha_2)\hat{\boldsymbol{n}}$$

where the point in question makes angles of $-\alpha_1$ and $+\alpha_2$ with the ends.

Axial induction due to current loop For all loop elements, $\boldsymbol{R}$ is perpendicular to $\mathrm{d}\boldsymbol{l}$. All components of $\boldsymbol{B}$ not parallel to the horizontal axis will cancel for the full circuit so the only terms remaining are the horizontal contributions to $\boldsymbol{B}$, of the form $\mathrm{d}\boldsymbol{B}\sin\alpha$. If $\hat{\boldsymbol{n}}$ is a unit vector along the horizontal axis,

$$\mathrm{d}B_n=\frac{\mu_0 I}{4\pi}\frac{\mathrm{d}l\,R}{R^3}\sin\alpha$$

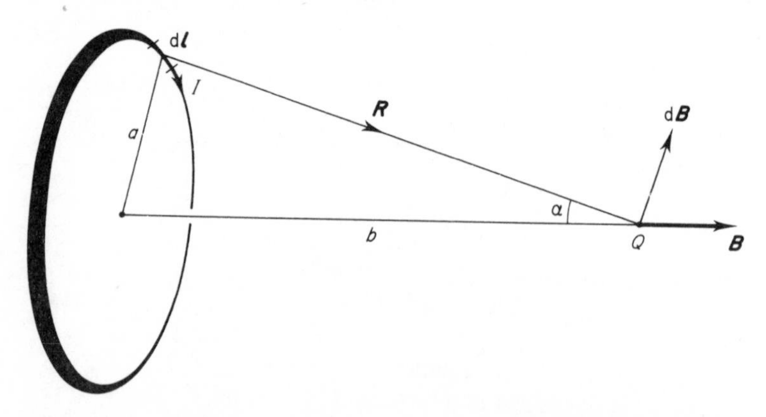

Fig. 3.4 Geometric construct needed to calculate B along the axis of a current loop. The result turns out to be independent of the shape of the loop provided that the point is far from the loop

so

$$\boldsymbol{B} = \frac{\mu_0 I}{4\pi} \frac{\sin \alpha}{R^2} \int \mathrm{d}l\, \hat{\boldsymbol{n}}.$$

Expressing R and $\sin \alpha$ in terms of the geometric quantities a and b

$$\boldsymbol{B} = \frac{\mu_0 I}{2} \frac{a^2}{(a^2 + b^2)^{3/2}} \hat{\boldsymbol{n}}. \tag{3.9}$$

At the centre of the loop, when $b = 0$, we find

$$\boldsymbol{B} = \frac{\mu_0 I}{2a} \hat{\boldsymbol{n}} \tag{3.10}$$

whilst at points along the axis but far away from the loop ($b^2 \gg a^2$)

$$\boldsymbol{B} = \frac{\mu_0}{4\pi} \frac{2IA}{b^3} \hat{\boldsymbol{n}}, \tag{3.11}$$

where the area A of the loop is small. For points far along the axis of an electric dipole of magnitude p, eq. (1.27) for the *electric* field becomes

$$E = \frac{1}{4\pi\epsilon_0} \frac{2p}{R^3}$$

or

$$\boldsymbol{E} = \frac{1}{4\pi\epsilon_0} \frac{2p}{R^3} \hat{\boldsymbol{n}}$$

and by analogy $IA\boldsymbol{n}$ is called the *magnetic dipole moment*, where $\boldsymbol{n}$ is normal to the plane of the loop, in the direction given by the right-hand screw rule for the current.

3.4 Ampère's circuital law

Equation (3.8) gives the magnetic induction due to a long wire carrying steady current I as

$$\boldsymbol{B} = \frac{\mu_0 I}{2\pi a} \hat{\boldsymbol{n}}$$

with the lines of constant $\boldsymbol{B}$ forming concentric rings around the wire. Consider the line integral $\oint_C \boldsymbol{B} \cdot \mathrm{d}\boldsymbol{l}$, where the path of integration C is a complete turn of any one of these concentric rings. $\boldsymbol{B}$ is constant along C and everywhere parallel to $\mathrm{d}\boldsymbol{l}$ hence

$$\oint_C \boldsymbol{B} \cdot \mathrm{d}\boldsymbol{l} = B \oint_C \mathrm{d}l$$

$$= 2\pi a B. \tag{3.12}$$

Substituting eq. (3.8) into eq. (3.12) gives

$$\oint \boldsymbol{B} \cdot \mathrm{d}\boldsymbol{l} = \mu_0 I \tag{3.13}$$

and it turns out that eq. (3.13) is true *in general* for any line integral taken around any closed curve C which encloses current I. The path need not coincide with a real, physical circuit and eq. (3.13) is usually referred to as *Ampère's circuital law*. It is the magnetostatic analogue of Gauss' theorem in electrostatics, and note that it offers a method of finding magnetic inductions due to circuits which have a high degree of symmetry, and where the line integral $\oint_C \boldsymbol{B} \cdot \mathrm{d}\boldsymbol{l}$ is easily evaluated analytically. Thus, consider the calculation of $\boldsymbol{B}$ for the infinite solenoid of Fig. 3.5(a). The flux lines are non-zero only *inside* the solenoid because contributions from individual loops cancel outside the solenoid and the flux lines are parallel to the sides (real solenoids have flux lines outside as in Fig. 3.5(b)). Assuming that the wire turns of Fig. 3.5(a) are very closely wound with N turns per unit length, then the current appearing in Ampère's law is the total current enclosed by the path of integration, so for a path chosen as in Fig. 3.6(a)

$$\oint_C \boldsymbol{B} \cdot \mathrm{d}\boldsymbol{l} = NL\, \mu_0 I. \tag{3.14}$$

A sign convention is implicit in the direction of travel around the path C: the

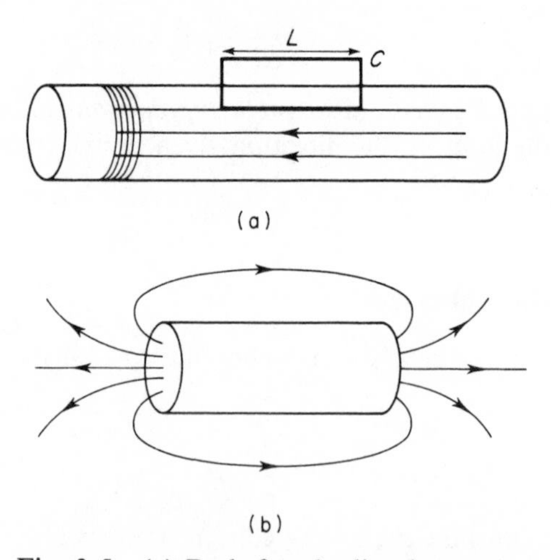

Fig. 3.5 (a) Path for the line integral in Ampère's circuital law calculation of B for an infinite solenoid. There are N turns of wire per unit length and the wire carries a steady current I. (b) A real solenoid. Flux lines no longer start and end at infinity

direction to make the line integral positive is determined by the right-hand screw rule for the current, and so eq. (3.14) becomes

$$BL = NL\,\mu_0 I$$

hence

$$B = \mu_0 NI. \tag{3.15}$$

Because the path of integration could have been chosen at any depth inside the solenoid, $\boldsymbol{B}$ is constant inside an infinite solenoid

3.5 The volume current density

Just as in electrostatics it proved useful to express Gauss' law $\oint \boldsymbol{E}\cdot \mathrm{d}\boldsymbol{S} = q/\epsilon_0$ in the differential form

$$\nabla \cdot \boldsymbol{E} = \varrho(\boldsymbol{r})/\epsilon_0$$

relating the derivatives of $\boldsymbol{E}$ at field point $\boldsymbol{r}$ to the local charge density $\varrho(\boldsymbol{r})$, so in magnetostatics a corresponding differential expression for eq. (3.13) exists relating the derivatives of $\boldsymbol{B}$ to a local *volume current density* $\boldsymbol{J}(\boldsymbol{r})$, as follows. Suppose that charge is flowing through the surface S of Fig. 3.6 then the direction of $\boldsymbol{J}$ is that of the direction of flow of charge, and its magnitude is the current per unit area:

$$I = \left(\frac{\mathrm{d}q}{\mathrm{d}t}\right)_{\text{through } S} = \int \boldsymbol{J}\cdot \mathrm{d}\boldsymbol{S}.$$

If $\mathrm{d}\tau$ is the volume of Fig. 3.6

$$I\,\mathrm{d}l = J\,\mathrm{d}S\cos\theta\,\mathrm{d}l$$

so

$$I\,\mathrm{d}\boldsymbol{l} = \boldsymbol{J}\,\mathrm{d}\tau. \tag{3.16}$$

The Biot–Savart law can be written

$$\boldsymbol{B} = \frac{\mu_0}{4\pi}\oint \frac{\boldsymbol{J}\times \boldsymbol{R}}{R^3}\,\mathrm{d}\tau \tag{3.17}$$

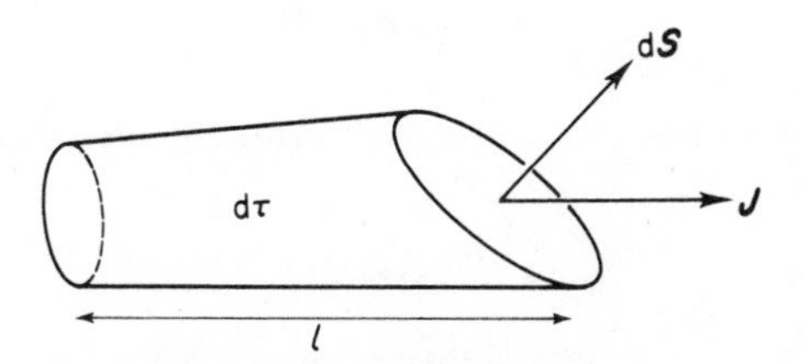

Fig. 3.6 Construct needed to demonstrate the relationship between current I and current density $\boldsymbol{J}\cdot \mathrm{d}\boldsymbol{S}$ in a surface element

whilst Ampère's law becomes

$$\oint \boldsymbol{B}\cdot \mathrm{d}\boldsymbol{l} = \mu_0 \int \boldsymbol{J}\cdot \mathrm{d}\boldsymbol{S} \tag{3.18}$$

and using Stokes' theorem of Appendix A to relate the line integral appearing in eq. (3.18) to a surface integral,

$$\int (\nabla \times \boldsymbol{B})\cdot \mathrm{d}\boldsymbol{S} = \mu_0 \int \boldsymbol{J}\cdot \mathrm{d}\boldsymbol{S}. \tag{3.19}$$

Because eq. (3.19) was derived without reference to the specific form of the surface S it must be true for any surface and hence the integrands must be equal. Hence

$$\nabla \times \boldsymbol{B} = \mu_0 \boldsymbol{J} \tag{3.20}$$

which is the differential form of Ampère's circuital law, corresponding to the differential form of Gauss' law

$$\nabla \cdot \boldsymbol{E} = \varrho/\epsilon_0.$$

3.6 The divergence of *B*

We first give an intuitive physical proof that $\nabla \cdot \boldsymbol{B} = 0$, then a direct mathematical derivation starting from the Biot–Savart law. As we noted earlier, there is a fundamental difference between magnetostatic and electrostatic fields in that magnetostatic flux lines are invariably closed, consistent with the view that magnetic monopoles do not exist. Consider a closed surface S drawn in an arbitrary magnetic induction $\boldsymbol{B}$. Since lines of $\boldsymbol{B}$ are closed, as much magnetic flux enters the surface as leaves, so

$$\oint \boldsymbol{B}\cdot \mathrm{d}\boldsymbol{S} = 0 \tag{3.21}$$

and using the divergence theorem of Appendix A to convert the surface integral into a volume integral

$$\int \nabla \cdot \boldsymbol{B}\, \mathrm{d}\tau = 0.$$

The integrand is also zero, because the argument has made no assumption about the surface, hence

$$\nabla \cdot \boldsymbol{B} = 0. \tag{3.22}$$

The indirect evidence for the validity of eq. (3.22) is immense. No magnetic monopoles have been observed, and since electromagnetic theory invariably gives answers in agreement with experiment any violations of eq. (3.22) must occur under exceptional circumstances. The development of the theory of electromagnetism depends on this negative observation.

A rigorous derivation of eq. (3.22), assuming the truth of the Biot–Savart law, is now given

$$B_a(r) = \frac{\mu_0}{4\pi}\oint_{C_a}\frac{dl_a \times (r - r_a)}{|r - r_a|^3}. \tag{3.23}$$

It is necessary to realize that the variable point of eq. (3.23) is the field point r, and we therefore wish to prove that

$$\nabla \cdot B_a(r) = 0,$$

where

$$\nabla \equiv \frac{\partial}{\partial x}\,i + \frac{\partial}{\partial y}\,j + \frac{\partial}{\partial z}\,k$$

acts on the coordinates of the field point r, *not* on the point r_a. Now

$$\begin{aligned}\nabla \cdot B_a &= \frac{\mu_0}{4\pi}\,\nabla \cdot \oint_{C_a}\frac{dl_a \times (r - r_a)}{|r - r_a|^3}\\ &= \frac{\mu_0}{4\pi}\oint_{C_a}\nabla \cdot \frac{dl_a \times (r - r_a)}{|r - r_a|^3}\\ &= \frac{\mu_0}{4\pi}\oint_{C_a}\left[\frac{(r - r_a)}{|r - r_a|^3}\cdot \nabla \times dl_a - dl_a \cdot \nabla \times \frac{(r - r_a)}{|r - r_a|^3}\right] \qquad (3.24)\end{aligned}$$

by a vector identity. Now $\nabla \times dl_a = 0$ in a view of the remarks above, and since

$$\nabla \times \frac{R}{R^3} \equiv -\nabla \times \nabla\left(\frac{1}{R}\right) \equiv 0$$

eq. (3.24) yields

$$\nabla \cdot B_a = 0$$

as required.

3.7 The vector potential

In electrostatics the vector field $E(r)$ turned out to be a conservative field and it was hence possible to introduce the useful scalar field $V(r)$. By contrast, Ampère's circuital law, eq. (3.13),

$$\oint B \cdot dl = \mu_0 I$$

shows that B is not in general conservative and so it is not in general useful to seek a scalar field $\psi(r)$ such that $B = -\nabla\psi$, because $\psi(r)$ would not have properties analogous to V. The exception is where the region of space of interest does not enclose any current and in such a case it can prove useful to

Table 3.1

Electrostatics	Magnetostatics
$\nabla \cdot \boldsymbol{E} = \varrho/\epsilon_0$	$\nabla \cdot \boldsymbol{B} = 0$
$\nabla \times \boldsymbol{E} = \boldsymbol{0}$	$\nabla \times \boldsymbol{B} = \mu_0 \boldsymbol{J}$

work with a *magnetostatic scalar potential.* In the general case, however, it proves useful to construct a different potential, the *vector potential* $\boldsymbol{A}$. The reason for introducing $\boldsymbol{A}$ is that it aids calculation of $\boldsymbol{B}$ in the magnetostatic case just as V aided calculation of $\boldsymbol{E}$ in the electrostatic one.

At the deep level of quantum mechanics, $\boldsymbol{A}$ is a more fundamental field than $\boldsymbol{B}$ since it is $\boldsymbol{A}$ that appears in the hamiltonians describing charged particles in electromagnetic fields rather than $\boldsymbol{B}$ (see Chapter 13). Since $\nabla \cdot \boldsymbol{B} = 0$ it is possible to define a vector field $\boldsymbol{A}$ by

$$\boldsymbol{B} = \nabla \times \boldsymbol{A} \tag{3.25}$$

because of the vector identity div curl $\equiv 0$. As in the electrostatic case where V was undefined to within a constant, so it is possible to add any vector whose curl is zero to $\boldsymbol{A}$ and still satisfy eq. (3.25). In particular if ϕ is a differentiable scalar field, then both $\boldsymbol{A}$ and $\boldsymbol{A} + \nabla\phi$ satisfy eq. (3.25) because of the vector identity curl grad $\equiv 0$.

The differential equations of electrostatics and magnetostatics are summarized in Table 3.1.

As we will see, the equations for $\nabla \cdot \boldsymbol{E}$ and $\nabla \cdot \boldsymbol{B}$ are true in general even if $\boldsymbol{E}$ and $\boldsymbol{B}$ are not constant in time. The divergence equations are two of Maxwell's equations, but the curl equations need some modification, as we shall discuss in Chapters 4 and 5.

3.8 Magnetostatic analogues of Laplace's and Poisson's equations

Combining the differential form of Ampère's circuital law with the definition of $\boldsymbol{A}$ gives

$$\nabla \times (\nabla \times \boldsymbol{A}) = \mu_0 \boldsymbol{J}$$

and using the vector identity curl curl = grad div $- \nabla^2$

$$\nabla(\nabla \cdot \boldsymbol{A}) - \nabla^2 \boldsymbol{A} = \mu_0 \boldsymbol{J}. \tag{3.26}$$

Because $\boldsymbol{A}$ is undetermined to within a field whose curl equals zero we are free to make the special choice that $\nabla \cdot \boldsymbol{A} = 0$. This process is called choosing a *gauge*, and eq. (3.26) becomes

$$-\nabla^2 \boldsymbol{A} = \mu_0 \boldsymbol{J} \tag{3.27}$$

which is the magnetostatic analogue of Poisson's equation $-\nabla^2 V = \varrho/\epsilon_0$.

In regions which contain no current density, eq. (3.27) reduces to the magnetostatic equivalent of Laplace's equation

$$\nabla^2 \boldsymbol{A} = \boldsymbol{0}. \tag{3.28}$$

3.9 Explicit form of A

Just as in Chapter 1 we were able to show that the potential due to a charge distribution $\varrho(\boldsymbol{r})$ was

$$V(\boldsymbol{r}) = \frac{1}{4\pi\epsilon_0}\int \frac{\varrho(\boldsymbol{r})}{r}\,\mathrm{d}\tau$$

we now seek the corresponding magnetostatic expression for $\boldsymbol{A}(\boldsymbol{r})$ in terms of the current density $\boldsymbol{J}(\boldsymbol{r})$. The method is to manipulate the Biot–Savart law so as to obtain $\boldsymbol{B}$ as the curl of a certain vector. From the definition of $\boldsymbol{A}$, that vector will be $\boldsymbol{A}$. It is important to distinguish the variable point $\boldsymbol{r}$ from the fixed point $\boldsymbol{r}_\mathrm{a}$. We start from

$$\boldsymbol{B}_\mathrm{a} = \frac{\mu_0}{4\pi}\oint_{C_\mathrm{a}} \frac{\boldsymbol{J}(\boldsymbol{r}_\mathrm{a})\times(\boldsymbol{r}-\boldsymbol{r}_\mathrm{a})}{|\boldsymbol{r}-\boldsymbol{r}_\mathrm{a}|^3}\,\mathrm{d}\tau_\mathrm{a}$$

where

$$\frac{\boldsymbol{r}-\boldsymbol{r}_\mathrm{a}}{|\boldsymbol{r}-\boldsymbol{r}_\mathrm{a}|^3} = -\nabla\frac{1}{|\boldsymbol{r}-\boldsymbol{r}_\mathrm{a}|}$$

so that

$$\boldsymbol{B}_\mathrm{a} = -\frac{\mu_0}{4\pi}\oint \boldsymbol{J}\times\nabla\frac{1}{|\boldsymbol{r}-\boldsymbol{r}_\mathrm{a}|}\,\mathrm{d}\tau_\mathrm{a}.$$

Using the vector identity

$$\nabla\times\frac{\boldsymbol{J}}{|\boldsymbol{r}-\boldsymbol{r}_\mathrm{a}|} \equiv \frac{1}{|\boldsymbol{r}-\boldsymbol{r}_\mathrm{a}|}\nabla\times\boldsymbol{J} - \boldsymbol{J}\times\nabla\frac{1}{|\boldsymbol{r}-\boldsymbol{r}_\mathrm{a}|}$$

we find that

$$\boldsymbol{B} = \frac{\mu_0}{4\pi}\oint\left[\nabla\times\frac{\boldsymbol{J}}{|\boldsymbol{r}-\boldsymbol{r}_\mathrm{a}|} - \frac{1}{|\boldsymbol{r}-\boldsymbol{r}_\mathrm{a}|}\nabla\times\boldsymbol{J}\right]\mathrm{d}\tau.$$

The second term in the integrand is zero because ∇ operates only on the variable point $\boldsymbol{r}$ and so

$$\boldsymbol{A} = \frac{\mu_0}{4\pi}\oint\frac{\boldsymbol{J}}{|\boldsymbol{r}-\boldsymbol{r}_\mathrm{a}|}\,\mathrm{d}\tau_\mathrm{a} \tag{3.29}$$

or

$$\boldsymbol{A} = \frac{\mu_0}{4\pi}\int\frac{\boldsymbol{J}}{r}\,\mathrm{d}\tau. \tag{3.30}$$

Equation (3.30) can be equally well expressed in terms of current elements

$$A=\frac{\mu_0}{4\pi}\int\frac{I\,\mathrm{d}\boldsymbol{l}}{r}. \tag{3.31}$$

3.10 The Lorentz force

The basic magnetostatic force law, eq. (3.1) applies to two complete interacting circuits. The derived differential law involving current elements

$$\mathrm{d}\boldsymbol{F}=I\,\mathrm{d}\boldsymbol{l}\times\boldsymbol{B} \tag{3.32}$$

could be invalid, since alternative expressions might yield the same result upon integration, with terms cancelling. However, if the validity of eq. (3.32) is assumed it is possible to deduce an expression for the force on a single moving charge. This expression can and has been verified experimentally, for example by following the motion of an ion in a mass spectrometer. Starting from

$$\boldsymbol{F}=\int\boldsymbol{J}\times\boldsymbol{B}\,\mathrm{d}\tau$$

then if the current density $\boldsymbol{J}$ is due to a density N of particles each carrying charge q and moving with velocity $\boldsymbol{v}$, since current is defined as $\mathrm{d}q/\mathrm{d}t$ we have

$$\boldsymbol{F}=\int Nq\boldsymbol{V}\times\boldsymbol{B}\,\mathrm{d}\tau$$

and as $N\,\mathrm{d}\tau$ is the number of particles in $\mathrm{d}\tau$, the force on a single charge is

$$\boldsymbol{F}=q\boldsymbol{v}\times\boldsymbol{B}. \tag{3.33}$$

If in addition an electrostatic field $\boldsymbol{E}$ is present

$$\boldsymbol{F}=q(\boldsymbol{E}+\boldsymbol{v}\times\boldsymbol{B}) \tag{3.34}$$

and this is called the *Lorentz force*; solution of the differential eq. (3.34) determines the trajectory of the charged particles in the fields $\boldsymbol{E}$ and $\boldsymbol{B}$. Obviously, the precise equations of motion can only be solved when the forms of $\boldsymbol{E}$ and $\boldsymbol{B}$ are known, and in general there only exist analytical solutions of eq. (3.34) for very simple arrangements of $\boldsymbol{E}$ and $\boldsymbol{B}$ (cf. Examples 3.6, 3.8 and 3.9). In the general case it is necessary to numerically integrate eq. (3.34) to discover the trajectory.

The force $\boldsymbol{F}=m\,\mathrm{d}^2\boldsymbol{r}/\mathrm{d}t^2$ and if the particle's speed is significant with regard to the speed of light it is necessary to modify the m in the equation (cf. Example 3.10);

In the special case of *uniform* fields, it is possible to progress to a general solution of eq. (3.34) as follows. By convention we resolve all vectors into components parallel and perpendicular to $\boldsymbol{B}$. Thus, writing $\boldsymbol{X}_\parallel$ and $\boldsymbol{X}_\perp$ for the parallel and perpendicular components of a certain vector $\boldsymbol{X}$, it is clear that $\boldsymbol{X}=\boldsymbol{X}_\parallel+\boldsymbol{X}_\perp$.

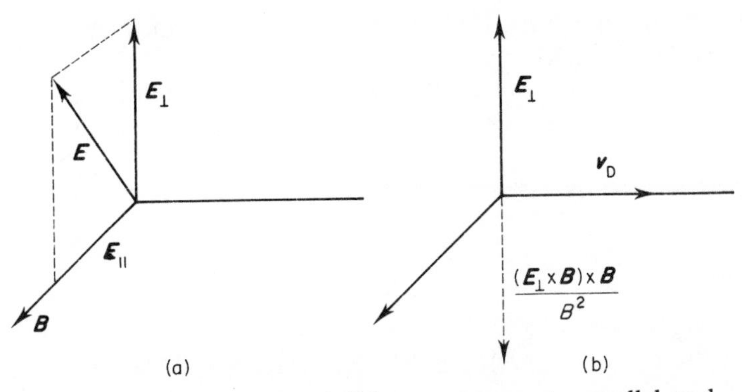

Fig. 3.7 (a) Resolution of $\boldsymbol{E}$ into components parallel and perpendicular to $\boldsymbol{B}$. (b) Relationship between the drift velocity $\boldsymbol{v}_\mathrm{D}$, $\boldsymbol{E}$ and $\boldsymbol{B}$

Equation (3.34) can thus be written

$$m\frac{\mathrm{d}}{\mathrm{d}t}(\boldsymbol{v}_\parallel + \boldsymbol{v}_\perp) = q(\boldsymbol{E}_\parallel + \boldsymbol{E}_\perp + (\boldsymbol{v}_\parallel + \boldsymbol{v}_\perp)\times \boldsymbol{B}) \tag{3.35}$$

and since $\boldsymbol{v}_\parallel \times \boldsymbol{B} = \boldsymbol{0}$ it is possible to separate (3.35) into the two equations

$$m\frac{\mathrm{d}\boldsymbol{v}_\parallel}{\mathrm{d}t} = q\boldsymbol{E}_\parallel \tag{3.36a}$$

$$m\frac{\mathrm{d}\boldsymbol{v}_\perp}{\mathrm{d}t} = q(\boldsymbol{E}_\perp + \boldsymbol{v}_\perp \times \boldsymbol{B}). \tag{3.36b}$$

Because we are dealing with uniform fields, the rate of change of $\boldsymbol{v}_\parallel$ is thus constant and in the direction of $\boldsymbol{v}_\parallel$, therefore this component of the motion is that of a body moving parallel to a uniform electric field.

Equation (3.36b) can be reduced to a simpler form, using the fact that

$$\frac{(\boldsymbol{E}_\perp \times \boldsymbol{B})\times \boldsymbol{B}}{B^2} = -\boldsymbol{E}_\perp$$

For the sake of argument, let us write

$$\boldsymbol{v}_\perp = \boldsymbol{u} + \frac{\boldsymbol{E}_\perp \times \boldsymbol{B}}{B^2} \tag{3.37}$$

Substitution of (3.37) into (3.36b) yields

$$m\frac{\mathrm{d}}{\mathrm{d}t}\left(\boldsymbol{u} + \frac{\boldsymbol{E}_\perp \times \boldsymbol{B}}{B^2}\right) = q(\boldsymbol{E}_\perp + \boldsymbol{u}\times \boldsymbol{B} - \boldsymbol{E}_\perp)$$

Now, $\boldsymbol{E}_\perp \times \boldsymbol{B}/B^2$ is a constant, so

$$m\frac{\mathrm{d}\boldsymbol{u}}{\mathrm{d}t} = q(\boldsymbol{u}\times \boldsymbol{B}) \tag{3.38}$$

The velocity $v_D = E \times B/B^2$ is called the *constant drift velocity* and its direction is indicated in Fig. 3.7.

Thus, to summarize: $v_\perp$ can be expressed as the sum of a constant drift velocity v_D and a variable velocity vector u, where u is found by the solution of eq. (3.38).

Examples

3.1 A square current loop of side $2a$ carries a steady current I. Use the Biot–Savart law (eq. (3.3)) to show that the magnetic induction at the centre of the loop has magnitude

$$\sqrt{2}\mu_0 I/\pi a$$

What is the direction of B?

3.2 Two parallel coils each of radius a are held with their faces parallel at a distance a apart. Each coil consists of N turns and carries a steady current I, which flows in the same direction in either coil. Use the Biot–Savart law to find B at points midway along the principal symmetry axis and show that B has magnitude

$$N\mu_0 I(0.8)^{3/2}/a$$

at a point midway between the coils. Show that the first and second derivatives of B are zero at this point.

Such an arrangement of coils can be used in practical applications where a magnetic induction is required that is uniform over a large volume.

3.3 Equation (3.9) gives the magnetic induction at a point distance b from a circular loop of wire carrying a steady current I. Use this result to establish that the magnetic induction at the centre of a long solenoid has magnitude

$$\mu_0 NI$$

where N is the number of turns of wire per unit length.

3.4 Use the method of Example 3.3 to show that the magnetic induction along the principal axis of a *short* solenoid is twice as strong at the centre of the solenoid as at the ends.

3.5 In a long cylindrical conductor of radius a, current flows parallel to the axis. The current density at a distance r from the axis is $(I/2ar)\cos(\pi r/a)$. Calculate the magnetic induction as a function of distance from the axis both inside and outside the conductor.

3.6 In the last few years, ion cyclotron resonance spectroscopy has been used to study a wide variety of chemical problems involving both isolated ions and ion–molecule reactions. An ion of charge Q and mass m moves with initial velocity v in a uniform magnetic induction $B = (0, 0, B_z)$. Show that the ion moves in a helix parallel to B and of radius mv_0/QB_z, where v_0 is the component of v perpendicular to B. What condition must be satisfied for the ion to describe a circle rather than a helix?

3.7 Electrons produced by a hot cathode and having negligible initial velocities are accelerated through a potential of 5 kV so as to form a narrow beam. Find the speed of the electrons after acceleration.

The beam, after acceleration, passes into a uniform magnetic induction initially at right angles to the beam. The beam is observed experimentally to be bent into a circle of radius 5 cm by the field. Calculate the magnetic flux density.

3.8 A particle of mass m and charge Q moves in a viscous medium under the influence of a uniform electric field $\boldsymbol{E}$, parallel to the x-axis. Show that the equations of motion can be written

$$\frac{\mathrm{d}\dot{x}}{\mathrm{d}t} + \dot{x}/\tau = qEm \qquad \left(\dot{x} \equiv \frac{\mathrm{d}x}{\mathrm{d}t}\right)$$

and that hence, for a particle starting at rest when $t = 0$

$$\dot{x} = (qE\tau/m)(1 - \exp(-t/\tau))$$

For an electron in a metal, the effect of collisions is similar to a viscous force and the value of τ is usually about 10^{-14} s. Thus the exponential term in the second equation quickly falls to zero and the velocity is proportional to the electric field strength. τ is known as the *relaxation time* since it gives a measure of the time required to reach the new equilibrium geometry when a field is altered.

3.9 A particle of charge Q and mass m starts from rest at the origin of coordinate in a uniform electric field $\boldsymbol{E}$ parallel to the x-axis and a uniform magnetic induction $\boldsymbol{B}$ parallel to the z-axis. Show that the coordinates of the particle at time t later will be

$$x = (E/\omega B)(1 - \cos \omega t)$$
$$y = -(E/\omega B)(\omega t - \sin \omega t)$$
$$z = 0$$

where $\omega = QB/m$.

Electrons are liberated with zero velocity from the negative plate of a parallel plate capacitor to which is applied an induction $\boldsymbol{B}$ parallel to the plates. Show that the electrons never reach the positive plate if the plate separation exceeds $2mE/eB^2$, where $\boldsymbol{E}$ is the field between the plates.

3.10 A particle of mass m and charge Q describes a circle of radius R in a plane perpendicular to the direction of a uniform magnetic induction $\boldsymbol{B}$.

(1) Show that

$$BQv = mv^2/R$$

(2) Show that the angular velocity ω is given by

$$\omega = BQ/m$$

This frequency is called the *cyclotron frequency* because it is the frequency at which an ion circulates in a cyclotron. Note that ω is independent of the velocity v of the particle. Strictly this is incorrect because the m appearing in the equation is itself a function of speed

$$m = m_0/(1 - v^2/c^2)$$

where c_0 is the speed of light in free space (see section 7.3).

(3) Calculate the cyclotron frequency for a deuteron in a field of 1 T assuming that the relativistic correction is not needed.

3.11 A steady current I flows in a long straight wire of length L. Show that, for points a distance r away from the wire such that $r \gg L$, the vector potential has magnitude

$$\frac{\mu_0 I}{4\pi} \log_e (2 + L^2/r^2)$$

What is the direction of $\boldsymbol{A}$?

Chapter 4

Magnetostatics in the presence of materials

4.1 Introduction

So far in the discussion of magnetostatics no allowance has been made for the presence of materials (other than the conductors carrying the currents). The aim of this chapter is to give an account of the behaviour of magnetic materials in external magnetic inductions. The *molecular* feature giving rise to permanent magnetism is angular momentum, but as far as the classical microscopic theory is concerned, neutral matter is equivalent to an assembly of magnetic dipole moments and these magnetic moments, when properly oriented, can produce a macroscopic magnetic moment. A body carrying a macroscopic magnetic dipole is said to be *magnetized*, and we are concerned here with the three main classes of magnetic materials, namely *diamagnetic*, *paramagnetic* and *ferromagnetic*. For the purposes of this chapter these classes can be distinguished by the behaviour of a sample of material in a non-uniform magnetic induction, e.g. near the pole pieces of a permanent magnet. A ferromagnetic material such as iron would be very strongly attracted into the field, a diamagnetic material such as glass or copper would be weakly repelled whilst a paramagnetic material such as manganese metal or liquid dioxygen O_2 would be pulled rather weakly into the field. Paramagnetism is a 'weak case of ferromagnetism'; there are special bulk features of ferromagnetic materials called *domains* caused by the very strong molecular magnetic dipoles which serve to make ferromagnetism so different from paramagnetism. The molecular properties of these materials will be dealt with in Chapter 10.

In Chapter 2 we discussed the polarization of dielectric media by external electric fields in terms of an induced electric dipole moment per unit volume

$$\mathrm{d}p = \boldsymbol{P}\,\mathrm{d}\tau,$$

where $\boldsymbol{P}$ is the electric polarization. In the same way magnetic materials are discussed by considering the induced magnetic dipole moment

$$\mathrm{d}\boldsymbol{m} = \boldsymbol{M}\,\mathrm{d}\tau, \tag{4.1}$$

where M is the *magnetization* of the sample. These are, however, significant differences between P and M: in electrostatics the effect of P is *always* to reduce the field inside the sample from the external field value. In the case of a paramagnetic sample the magnetization acts so as to *increase* the field, and the behaviour of ferromagnetic materials is even more complicated because samples retain their history of magnetization (they are said to display *hysteresis*).

As in the electrostatic case eq. (4.1) implies that $d\tau$ is sufficiently large on the molecular scale as to be homogeneous.

By analogy with the electrostatic case it might be reasonably expected that for a given material a functional relationship of the form

$$\boldsymbol{M} = \boldsymbol{M}(\boldsymbol{B})$$

would exist but for historical reasons it is first necessary to introduce a derived field H called the *magnetic field*, and then discuss the relationship

$$\boldsymbol{M} = \boldsymbol{M}(\boldsymbol{H}).$$

It is, however, important to realize that the derived field H is the magnetostatic equivalent of D whilst B is that of E. For a very long time, however, it was thought that H was the fundamental magnetostatic field.

4.2 Magnetization current densities

In Chapter 2 we showed that, as far as the calculation of effects outside a dielectric was concerned, the electric polarization P was formally equivalent to certain bound volume and surface charges. We now show that a corresponding result holds for magnetic materials, but it must be stressed that both P and M are experimentally determined quantities that have to be measured for a given sample at the field strengths of interest. In that respect the introduction of bound volume charges, etc. is purely a convenient mathematical device to aid, for example, the calculation of the potential outside a given dielectric of *known* polarization $P(E)$.

Thus one might expect that, just as the electrostatic potential can be interpreted as arising from certain *charges*, so the magnetostatic vector potential will have an interpretation in terms of certain *current densities*. The analogy is exact, the only difference in the derivation arising because the vector potential due to a magnetic dipole m,

$$\boldsymbol{A}(\boldsymbol{R}) = \frac{\mu_0}{4\pi} \boldsymbol{m} \times \frac{(\boldsymbol{R} - \boldsymbol{r})}{|\boldsymbol{R} - \boldsymbol{r}|^3}$$

involves a vector cross product whilst the potential due to an electric dipole p

$$V(\boldsymbol{R}) = \frac{1}{4\pi\epsilon_0} \boldsymbol{p} \cdot \frac{(\boldsymbol{R} - \boldsymbol{r})}{|\boldsymbol{R} - \boldsymbol{r}|^3}$$

involves a scalar product. Thus in detail, the magnetic moment per unit volume is

$$\mathrm{d}\boldsymbol{m} = \boldsymbol{M}\,\mathrm{d}\tau$$

and its contribution to $\boldsymbol{A}$ is

$$\mathrm{d}\boldsymbol{A} = \frac{\mu_0}{4\pi}\,\mathrm{d}\boldsymbol{m} \times \frac{(\boldsymbol{R}-\boldsymbol{r})}{|\boldsymbol{R}-\boldsymbol{r}|^3}. \tag{4.2}$$

As in the electrostatic case the following vector identities are substituted into eq. (4.2)

$$\frac{\boldsymbol{R}-\boldsymbol{r}}{|\boldsymbol{R}-\boldsymbol{r}|^3} = \nabla \left|\frac{1}{\boldsymbol{R}-\boldsymbol{r}}\right| \tag{4.3}$$

and

$$\boldsymbol{M}\times \nabla \left|\frac{1}{\boldsymbol{R}-\boldsymbol{r}}\right| = \frac{\nabla \times \boldsymbol{M}}{|\boldsymbol{R}-\boldsymbol{r}|} - \nabla \times \frac{\boldsymbol{M}}{|\boldsymbol{R}-\boldsymbol{r}|} \tag{4.4}$$

and using Stokes' theorem to relate a surface to a volume integral yields

$$\boldsymbol{A}(\boldsymbol{R}) = \frac{\mu_0}{4\pi}\int_V \frac{\nabla \times \boldsymbol{M}}{|\boldsymbol{R}-\boldsymbol{r}|}\,\mathrm{d}\tau + \frac{\mu_0}{4\pi}\oint_S \frac{\boldsymbol{M}\times \boldsymbol{n}}{|\boldsymbol{R}-\boldsymbol{r}|}\,\mathrm{d}S, \tag{4.5}$$

the surface integral being over the surface S bounding the volume V of the magnetized body, $\boldsymbol{n}$ being a unit normal. Thus eq. (4.5) can be given the interpretation that the vector potential due to a magnetized body is formally due to a *volume current density*

$$\boldsymbol{J}_{\mathrm{M}} = \nabla \times \boldsymbol{M} \tag{4.6}$$

and a *surface current density*

$$\boldsymbol{\lambda}_{\mathrm{M}} = \boldsymbol{M} \times \boldsymbol{n} \tag{4.7}$$

which are often called the *Ampèrian* volume and surface current densities or *bound* current densities. It is also useful to give a direct physical picture of the surface current: consider Fig. 4.1.

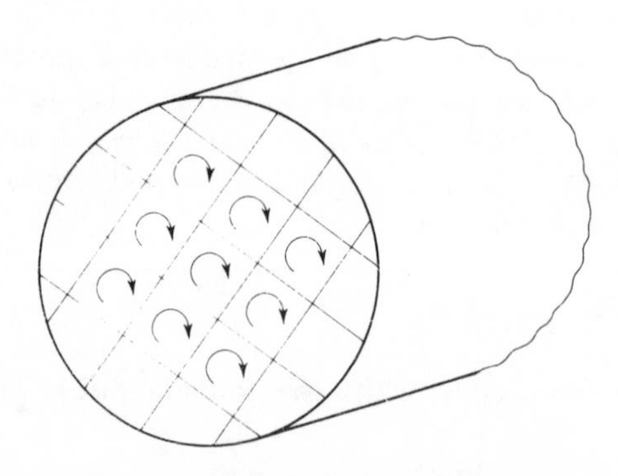

Fig. 4.1 Surface current density

At points in the interior volume shown of a uniformly magnetized material, the magnetization current is zero because the current associated with one dipole will be cancelled by that associated with a neighbouring dipole. However, at the surface such a cancellation does not occur and the net result is a current λ_M circulating on the surface.

If $\boldsymbol{M}$ is not uniform then there will be no longer a complete cancellation internally and there will be a resultant current inside the material.

4.3 The magnetic field H

Ampère's circuital law

$$\oint \boldsymbol{B}\cdot \mathrm{d}\boldsymbol{l} = \mu_0 I_{\text{total}}$$

and its equivalent differential form

$$\nabla \times \boldsymbol{B} = \mu_0 \boldsymbol{J}_{\text{total}}$$

refer to the *total* current and charge densities, I_{total} and $\boldsymbol{J}_{\text{total}}$: as we have just seen, magnetized material is formally equivalent to certain current densities, so

$$I_{\text{total}} = I_{\text{free}} + I_M$$

$$\boldsymbol{J}_{\text{total}} = \boldsymbol{J}_{\text{free}} + \boldsymbol{J}_M \tag{4.8}$$

where $\boldsymbol{J}_M = \nabla \times \boldsymbol{M}$. As in the electrostatic case it is often useful to modify Ampère's law so that only free currents appear on the right-hand side. Now

$$\nabla \times \boldsymbol{B} = \mu_0 \boldsymbol{J}_{\text{total}} \tag{4.9}$$

$$\nabla \times \boldsymbol{B} = \mu_0(\boldsymbol{J}_{\text{free}} + \boldsymbol{J}_M) \tag{4.10}$$

and since $\boldsymbol{J}_M = \nabla \times \boldsymbol{M}$ we find on rearranging eq. (4.10)

$$\nabla \times (\boldsymbol{B} - \mu_0 \boldsymbol{M}) = \mu_0 \boldsymbol{J}_{\text{free}} \tag{4.11}$$

which is written for historical reasons as

$$\nabla \times \left(\frac{\boldsymbol{B}}{\mu_0} - \boldsymbol{M}\right) = \boldsymbol{J}_{\text{free}}. \tag{4.12}$$

Equation (4.12) serves to define the *magnetic field* $\boldsymbol{H}$ as

$$\boldsymbol{H} = \frac{\boldsymbol{B}}{\mu_0} - \boldsymbol{M}. \tag{4.13}$$

Equation (4.13) is to be compared to eq. (2.18) for the electric displacement

$$\boldsymbol{D} = \epsilon_0 \boldsymbol{E} + \boldsymbol{P}$$

The reason for the 'unsymmetrical' treatment of electrostatics and magnetostatics is historical: it was thought that $\boldsymbol{H}$ was the fundamental magnetic vector field rather than $\boldsymbol{B}$. The proper association is

$$\boldsymbol{B} \leftrightarrow \boldsymbol{E}$$

$$\boldsymbol{H} \leftrightarrow \boldsymbol{D} \tag{4.14}$$

and on the molecular scale it is $\boldsymbol{B}$ and $\boldsymbol{E}$ which give the forces on charges. $\boldsymbol{H}$ and $\boldsymbol{D}$ are defined as convenient composites which describe the properties of macroscopic systems compactly.

4.4 Ampère's circuital law

When written in terms of the magnetic field $\boldsymbol{H}$, Ampère's circuital law becomes

$$\oint \boldsymbol{H} \cdot \mathrm{d}\boldsymbol{l} = I_{\text{free}} \tag{4.15}$$

or, in its differential form

$$\nabla \times \boldsymbol{H} = \boldsymbol{J}_{\text{free}}. \tag{4.16}$$

These equations (4.15) and (4.16) are a more general form of Ampère's law in that they can be used to calculate the magnetic field in the presence of magnetic materials.

4.5 Magnetic susceptibility

As in the case of dielectrics, there exists a functional relationship $\boldsymbol{M} = \boldsymbol{M}(\boldsymbol{B})$, but it is more usual to investigate the dependence of $\boldsymbol{M}$ on $\boldsymbol{H}$. If $\boldsymbol{M} \neq \boldsymbol{0}$ when $\boldsymbol{H} = \boldsymbol{0}$ then the material is magnetized, or has *permanent magnetization*. Many materials for which $\boldsymbol{M}(\boldsymbol{H} = \boldsymbol{0}) \neq \boldsymbol{0}$ have a very non-linear dependence of $\boldsymbol{M}$ on $\boldsymbol{H}$, and the relationship need not be a single-valued one. The simplest possible case, however, is that of a linear isotropic homogeneous magnetic material, where

$$\boldsymbol{M} = \chi_{\mathrm{m}} \boldsymbol{H} \tag{4.17}$$

and χ_{m} is called the *magnetic susceptibility*. Generally χ_{m} is very small but unlike the electrostatic case χ_{m} can be positive or negative. *All* materials have a diamagnetic (negative) contribution to their susceptibilities but in paramagnetic materials the paramagnetic contribution dominates. This behaviour is discussed fully in Chapter 10 from the molecular standpoint.

Combining eqs (4.13) and (4.17) yields

$$\boldsymbol{B} = \mu_0(1 + \chi_{\mathrm{m}})\boldsymbol{H} = \mu_{\mathrm{r}}\mu_0\boldsymbol{H} = \mu\boldsymbol{H} \tag{4.18}$$

where μ_{r} is the *relative permeability* and μ the *permeability*. Both μ_{r} and χ_{m} are dimensionless. Note, however, that eq. (4.18) is not valid *in general*: it is *not* a fundamental equation of electromagnetism. Equations such as $\boldsymbol{B} = \mu\boldsymbol{H}$ and $\boldsymbol{D} = \epsilon\boldsymbol{E}$ are called *constitutive equations* and are necessary in order to make progress in solving the basic electromagnetic equations in the presence of matter.

4.6 Magnetostatic energy

In Section 2.7 we showed that the stored energy per unit volume of a dielectric medium is

$$\tfrac{1}{2}\boldsymbol{D}\cdot\boldsymbol{E}.$$

We derived this general result by initially considering the special case of the work required to charge up a parallel-plate capacitor. We might expect by analogy that the energy density associated with a magnetostatic field is

$$\tfrac{1}{2}\boldsymbol{B}\cdot\boldsymbol{H} \tag{4.19}$$

and this is indeed the case. A general proof starts with the expression for the magnetostatic energy of a system of currents

$$U=\frac{1}{2}\int \boldsymbol{J}_{\mathrm{f}}\cdot\boldsymbol{A}\,\mathrm{d}\tau, \tag{4.20}$$

where $\boldsymbol{J}_{\mathrm{f}} = \nabla \times \boldsymbol{H}$. Putting

$$\boldsymbol{A}\cdot(\nabla\times\boldsymbol{H}) = \boldsymbol{H}.(\nabla\times\boldsymbol{A}) - \nabla\cdot(\boldsymbol{A}\times\boldsymbol{H})$$

and dropping zero terms from the integral yields eq. (4.19).

4.7 Summary

It is convenient at this stage to summarize the basic equations of electrostatics and magnetostatics, in order to compare corresponding equations. This is done in Table 4.1.

An obvious asymmetry between the basic force laws is that the electrostatic proportionality constant has ϵ_0 in the denominator whilst μ_0 appears as numerator in the magnetostatic case. The equations

$$\nabla\cdot\boldsymbol{E} = \varrho/\epsilon_0 \qquad \nabla\cdot\boldsymbol{B} = 0$$

are two equations which prove to be valid generally even if the charges are not at rest and the currents are not steady ones. The corresponding equations for $\nabla\times\boldsymbol{B}$ and $\nabla\times\boldsymbol{E}$ are not yet sufficiently general. We show in Chapters 5 and 6 how they need to be modified to take account of such dynamic cases. The four equations for $\nabla\cdot\boldsymbol{E}$, $\nabla\cdot\boldsymbol{B}$, $\nabla\times\boldsymbol{E}$ and $\nabla\times\boldsymbol{B}$ constitute *Maxwell's equations*, the basic equations of electromagnetism *in vacuo*. When matter is present, constitutive equations such as $\boldsymbol{D} = \epsilon\boldsymbol{E}$ need to be added to Maxwell's equations in order to permit their solution.

Table 4.1 Summary of electrostatic and magnetostatic formulae

	Electrostatics	Magnetostatics
1. *Force law*	$\frac{1}{4\pi\epsilon_0}\iint \frac{\varrho_a(\boldsymbol{r}_a)\,\varrho_b(\boldsymbol{r}_b)\boldsymbol{r}_{ab}}{r_{ab}^3}\,d\tau_a\,d\tau_b$	$\frac{\mu_0}{4\pi}\iint \frac{\boldsymbol{J}_a(\boldsymbol{r}_a)\times[\boldsymbol{J}_b(\boldsymbol{r}_b)\times\boldsymbol{r}_{ab}]}{r_{ab}^3}\,d\tau_a\,d\tau_b$
2. *Field equation*	$\boldsymbol{E}=\frac{1}{4\pi\epsilon_0}\int\frac{\varrho(\boldsymbol{R})\boldsymbol{R}\,d\tau}{R^3}$	$\boldsymbol{B}=\frac{\mu_0}{4\pi}\int\frac{\boldsymbol{J}(\boldsymbol{R})\times\boldsymbol{R}}{R^3}\,d\tau$
3. *Potential*	$V=\frac{1}{4\pi\epsilon_0}\int\frac{\varrho(\boldsymbol{R})}{R}\,d\tau$	$\boldsymbol{A}=\frac{\mu_0}{4\pi}\int\frac{\boldsymbol{J}(\boldsymbol{R})}{R}\,d\tau$
4. *Potential due to dipole*	$V=\frac{\mathrm{p}_e\cdot\boldsymbol{R}}{4\pi\epsilon_0 R^3}$	$\boldsymbol{A}=\frac{\mu_0}{4\pi}\frac{\boldsymbol{m}\times\boldsymbol{R}}{R^3}$
5. *Definition of potential*	$\boldsymbol{E}=-\nabla V$	$\boldsymbol{B}=\nabla\times\boldsymbol{A}$
6. *Special properties of field in static cases*	$\nabla\cdot\boldsymbol{E}=\varrho/\epsilon_0$	$\nabla\cdot\boldsymbol{B}=0$
	$\nabla\times\boldsymbol{E}=\boldsymbol{0}$	$\nabla\times\boldsymbol{B}=\mu_0\boldsymbol{J}$
7. *Materials*	$\boldsymbol{D}=\epsilon_0\boldsymbol{E}+\boldsymbol{P}$	$\boldsymbol{H}=\frac{\boldsymbol{B}}{\mu_0}+\boldsymbol{M}$
	$\nabla\cdot\boldsymbol{D}=\varrho_{\text{free}}$	$\nabla\times\boldsymbol{H}=\boldsymbol{J}_{\text{free}}$
	$\varrho_p=-\nabla\cdot\boldsymbol{P}$	$\boldsymbol{J}_M=\nabla\times\boldsymbol{M}$
	$\sigma_p=\boldsymbol{P}\cdot\boldsymbol{n}$	$\lambda_M=\boldsymbol{M}\times\boldsymbol{n}$
8. *Energy density*	$\frac{1}{2}\boldsymbol{D}\cdot\boldsymbol{E}$	$\frac{1}{2}\boldsymbol{B}\cdot\boldsymbol{H}$

Examples

4.1 A cube of side a has a magnetization given by

$$\boldsymbol{M}=\left(-\frac{M}{a}\right)y\hat{\boldsymbol{x}}+\left(\frac{M}{a}\right)x\hat{\boldsymbol{y}}$$

where M is a constant. Show that

$$J_{\mathrm{M}} = \frac{2M}{a} \hat{z}$$

4.2 A sphere of radius a has its centre at the coordinate origin. Its magnetization, which is non-uniform, is given by

$$M = (\alpha z^2 + \beta)\hat{z}$$

where α and β are constants. Find the magnetization current densities J_{M} and λ_{M} in polar coordinates.

Chapter 5

Induced e.m.f

5.1 Introduction

Chapter 3 dealt with magneto*statics*, the magnetic effects produced by steady currents (i.e., when the *fields* were *static*.). In this chapter we examine the effects of currents that vary with time. Such changing currents give rise to varying magnetic inductions and we will see that under such conditions the electric field $\boldsymbol{E}$ is no longer a conservative field, and so $\nabla \times \boldsymbol{E} \neq \boldsymbol{0}$. We will also discover how the equation for $\nabla \times \boldsymbol{E}$ can be modified to account for the changing magnetic induction and hence arrive at the third basic equation of electromagnetism.

5.2 Magnetic flux

In Chapter 3 we saw that $\oint_S \boldsymbol{B} \cdot \mathrm{d}\boldsymbol{S} = 0$, whenever the surface integral is taken over a *closed* surface S. The result can be stated equivalently that either $\nabla \cdot \boldsymbol{B} = 0$ or that magnetic monopoles do not exist, each statement asserting the same fundamental law of electromagnetism. In general we may be interested in finding the *magnetic flux* Φ through an arbitrary surface which is not necessarily a closed one. The magnetic flux through surface S is defined by

$$\Phi = \int_S \boldsymbol{B} \cdot \mathrm{d}\boldsymbol{S} \tag{5.1}$$

and if the surface is closed, $\Phi = 0$. Magnetic flux is measured in webers, Wb, and inspection of eq. (5.1) shows that $\mathrm{Wb} = \mathrm{T\ m}^2$. The magnetic induction $\boldsymbol{B}$ thus has units $\mathrm{Wb\ m}^{-2}$ and is often referred to as magnetic flux density. From the definition of the vector potential $\boldsymbol{B} = \nabla \times \boldsymbol{A}$, eq. (5.1) can be rewritten

$$\Phi = \int_S (\nabla \times \boldsymbol{A}) \cdot \mathrm{d}\boldsymbol{S}$$

or using Stokes' theorem to relate the surface integral to a line integral

$$\Phi = \oint_C \boldsymbol{A} \cdot \mathrm{d}\boldsymbol{l}, \tag{5.2}$$

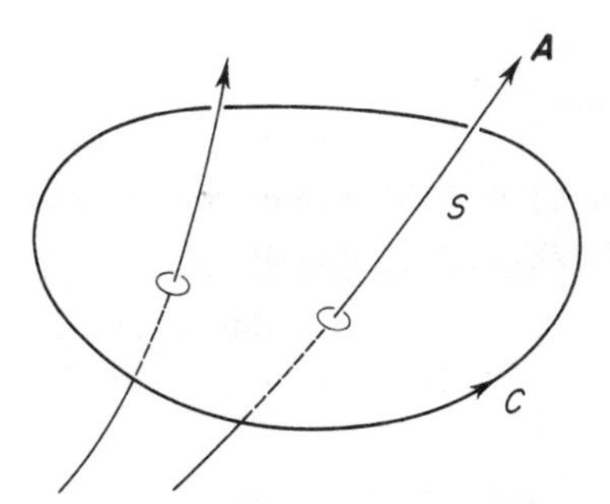

Fig. 5.1 The flux of vector field through surface S is given by the surface integral $\int \boldsymbol{A} \cdot d\boldsymbol{S}$. According to Stokes' theorem this is equal to $\oint \boldsymbol{A} \cdot d\boldsymbol{l}$ where the line integral is taken around the path C in the direction indicated

where the line integral is taken over any closed curve bounding the surface S. The flux through S can be either positive or negative, and the sign is taken in accordance with the familiar right-hand screw convention illustrated in Fig 5.1.

5.3 Faraday's law

If the circuit C of Fig. 5.2 does not contain any batteries or other sources of electromotive forces (e.m.f.) then the magnetic flux passing through the circuit is constant and it is observed experimentally that no current flows in the circuit. Faraday showed that an *induced current* could be produced in the circuit by switching on and switching off the magnetic flux. It is usual to discuss the effect in terms of the *induced e.m.f.* in the circuit.

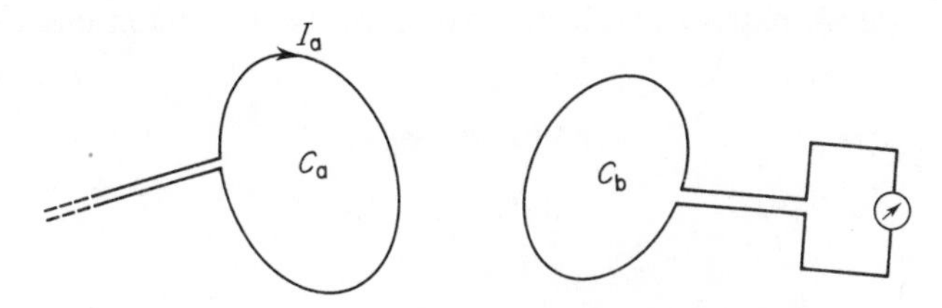

Fig. 5.2 A current I_a is switched into circuit C_a, causing an e.m.f. to be detected in circuit C_b. The e.m.f. is only detected for the time when the current in C_a is varying

Faraday showed that:

(a) When the magnetic flux Φ_a due to circuit A passing through (or *linking*) circuit C was changing with time, an e.m.f. ϵ_b was induced in circuit B.

(b)
$$\epsilon_b \propto d\Phi_a/dt \tag{5.3}$$

The full result is

$$\epsilon_b = -\frac{d\Phi_a}{dt} \tag{5.4}$$

the negative sign (Lenz' law) representing the sense of the induced e.m.f. compared to the direction defined by Fig. 5.1: the induced e.m.f. acts so as to cause currents in C whose magnetic effect oppose those of the original flux. Lenz' law is thus analogous to Le Chatelier's principle in thermodynamics which states that the response of a system in stable equilibrium to external effects altering that state of equilibrium is in the direction as to oppose the change. We note that eq. (5.4) relates electric and magnetic effects and that the use of a coherent four-dimensional set of units (SI) means that no dimensional factors need to be introduced.

In accord with the general philosophy of expressing electromagnetic phenomena in field form we rewrite eq. (5.4) in terms of the magnetic induction $\boldsymbol{B}_a$ due to circuit C_a. From eq. (5.1)

$$\epsilon_b = -\frac{d}{dt}\int_b \boldsymbol{B}_a \cdot d\boldsymbol{S}_b \tag{5.5}$$

and because the area is constant with time the order of differentiation and integration can be reversed to give

$$\epsilon_b = -\int_b \frac{\partial \boldsymbol{B}_a}{\partial t} \cdot d\boldsymbol{S}_b. \tag{5.6}$$

In Chapter 1 we found that electrostatic fields are conservative fields, i.e. $\oint \boldsymbol{E} \cdot d\boldsymbol{l} = 0$. However, if batteries are present in the integration path this is no longer the case. A battery is said to have e.m.f. ϵ if it maintains a potential difference

$$\begin{aligned} \epsilon &= V(T_2) - V(T_1) \\ &= \int_{T_1}^{T_2} \boldsymbol{E} \cdot d\boldsymbol{l} \end{aligned} \tag{5.7}$$

across its terminals T_1 and T_2. Faraday's experiments demonstrated that an e.m.f. can exist in a closed circuit even when no batteries are present. Combining eqs (5.7) and (5.6) we find

$$\int \boldsymbol{E} \cdot d\boldsymbol{l} + \oint \frac{\partial \boldsymbol{B}}{\partial t} \cdot d\boldsymbol{S} = 0. \tag{5.8}$$

Using Stokes' theorem to relate the line integral to a surface integral,

$$\int_S (\nabla \times \boldsymbol{E}) \cdot \mathrm{d}\boldsymbol{S} + \int_S \frac{\partial \boldsymbol{B}}{\partial t} \cdot \mathrm{d}\boldsymbol{S} = 0 \tag{5.9}$$

and because no assumptions have been made about the surface S the integrand must be zero

$$\nabla \times \boldsymbol{E} + \frac{\partial \boldsymbol{B}}{\partial t} = 0. \tag{5.10}$$

Thus eq. (5.10) is the first link we have discovered between the phenomena of magnetism and electricity.

5.4 The scalar potential

Equation (5.10) shows clearly that in the general case $\boldsymbol{E}$ is not a conservative field: $\boldsymbol{E}$ is conservative only in the electrostatic case. It proves desirable to manipulate eq. (5.10) in order to find a vector with zero curl: this curl-free vector is then a conservative field and a scalar potential can be introduced. From the definition of $\boldsymbol{A}$, eq. (5.10) becomes

$$\nabla \times \left(\boldsymbol{E} + \frac{\partial \boldsymbol{A}}{\partial t}\right) = 0 \tag{5.11}$$

and so $\boldsymbol{E} + \partial \boldsymbol{A}/\partial t$ is the desired conservative vector field. It is helpful to redefine the scalar potential V such that

$$\boldsymbol{E} + \frac{\partial \boldsymbol{A}}{\partial t} = -\nabla V \tag{5.12}$$

showing that in the general case $\boldsymbol{E}$ depends on both $\boldsymbol{A}$ and V. In the static case $\partial \boldsymbol{A}/\partial t = 0$ and eq. (5.12) reduces to the familiar $\boldsymbol{E} = -\nabla V$ of Chapter 1.

5.5 Mutual inductance

If the magnetic flux from circuit C_a of Fig 5.2 links circuit C_b, Faraday's law shows how changing the current I_a in circuit C_a will induce an e.m.f. ϵ_b in circuit C_b. This induced e.m.f. is linked to the varying current by the *mutual inductance* $M_{\mathrm{a,b}}$ of the two circuits such that

$$\epsilon_\mathrm{b} = -M_{\mathrm{b,a}} \frac{\mathrm{d}I_\mathrm{a}}{\mathrm{d}t} \tag{5.13}$$

The negative sign reflects Lenz' law and we will see shortly that $M_{\mathrm{a,b}} = M_{\mathrm{b,a}}$ so that suffixes are unnecessary. The unit of mutual inductance is the henry, H, where H = Wb A^{-1}, so that a system in which an e.m.f. of 1 V is generated in one circuit by the current in a second varying at the rate of 1 As^{-1} has an inductance of 1 H.

To find an analytical expression for M, we start from eq. (5.2)

$$\Phi_b = \oint_b \boldsymbol{A}_a \cdot d\boldsymbol{l}_b$$

for the flux through circuit C_b of Fig. 5.2 due to circuit C_a. Using eq. (3.31) for $\boldsymbol{A}_a$

$$\Phi_b = \oint_b \left\{ \frac{\mu_0 I_a}{4\pi} \oint \frac{d\boldsymbol{l}_a}{r} \right\} \cdot d\boldsymbol{l}_b$$

$$= \frac{\mu_0 I_a}{4\pi} \oint_a \oint_b \frac{d\boldsymbol{l}_a \cdot d\boldsymbol{l}_b}{r}$$

whence

$$M = \frac{\mu_0}{4\pi} \oint_a \oint_b \frac{d\boldsymbol{l}_a \cdot d\boldsymbol{l}_b}{r} \tag{5.14}$$

This is known as the *Neumann* formula for M. It demonstrates clearly the symmetry of $M(M_{a,b} = M_{b,a})$, and it will be noted that mutual inductance is a purely geometric factor calculable in principle from the geometries and relative orientations of circuits C_a and C_b. When multiplied by the current in one circuit it gives the magnetic flux linking the other circuit.

5.6 Self inductance

A single circuit is inevitably linked by its own flux, and the *self-inductance* L is defined by

$$\Phi_a = L_a I_a \tag{5.15}$$

Neumann's formula eq. (5.14) becomes

$$L_a = \frac{\mu_0}{4\pi} \oint_a \oint_a \frac{d\boldsymbol{l}_a \cdot d\boldsymbol{l}'_a}{r} \tag{5.16}$$

where the double integral is taken twice over the circuit in question. L is always positive because of the sign conventions chosen. For example, we showed in Section 3.4 that the magnetic induction inside a long solenoid with N turns of wire per unit length is

$$B = \mu_0 NI/l$$

where l is the length of the solenoid. Thus

$$\Phi = \frac{\mu_0 NI}{l} \cdot \pi R^2$$

and the self-inductance is

$$L = \frac{N\Phi}{I} = \frac{\mu_0 N^2 I \, \pi R^2}{lI} = \frac{\mu_0 N^2 \, \pi R^2}{l}$$

There are, however, grave practical problems in calculating (5.16).

Example

5.1 Use the Biot–Savart law to calculate the magnetic induction B at the centre of a circular loop of radius r carrying a current I.

Suppose that the loop represents an aromatic π-system and that the current arises from a circulation of the six π-electrons induced by switching on a uniform magnetic induction $\boldsymbol{B}_0$ perpendicular to the loop. Given that Faraday's law of electromagnetic induction leads to the force on an electron

$$F = \frac{e}{2\pi r}\frac{\mathrm{d}\Phi}{\mathrm{d}t}$$

where $-e$ is the electronic charge and $\mathrm{d}\Phi/\mathrm{d}t$ is the rate of change of magnetic flux through the loop, write down the acceleration of an electron and obtain the total change in its velocity produced by the final magnetic induction B_0. Show that the total π-electron current so induced is $-3e^2B_0/2\pi m$, where m is the electron mass, and obtain the resultant magnetic induction at the centre of the loop in terms of $\boldsymbol{B}_0$. In 1,4 decamethylenebenzene, a chain of 10 CH_2 groups joins the *para*-positions of a benzene ring in such a way that the protons of the central methylene groups are close to the centre of the ring. Compare qualitatively the magnetic resonance frequency of those protons with that of the protons bonded directly to the benzene ring.

Chapter 6

Maxwell's equations

6.1 Introduction

So far we have found three of the four fundamental equations of electromagnetism. From our study of electrostatics in Chapter 1 we have

$$\nabla \cdot \boldsymbol{E} = \varrho/\epsilon_0 \tag{6.1}$$

whilst the corresponding equation for the divergence of $\boldsymbol{B}$ was shown in Chapter 3 to be

$$\nabla \cdot \boldsymbol{B} = 0 \tag{6.2}$$

In the case where the currents were not steady Chapter 5 revealed that

$$\nabla \times \boldsymbol{E} = -\frac{\partial \boldsymbol{B}}{\partial t} \tag{6.3}$$

and we have already found from Chapter 3 a corresponding equation for the curl of $\boldsymbol{B}$ for the special case of steady currents

$$\nabla \times \boldsymbol{B} = \mu_0 \boldsymbol{J} \tag{6.4}$$

In the same way that $\nabla \times \boldsymbol{E} = \boldsymbol{0}$ did not cater for the effects of varying currents, so eq. (6.4) is not completely general: it was Maxwell who showed how Ampère's law, eq. (6.4), needs to be modified to take account of *electric* fields that vary with time. A great triumph for Maxwell's discovery was the demonstration that the four equations could be solved with appropriate boundary conditions to give electromagnetic waves. *In vacuo* these waves travel at the speed of light, thus demonstrating the electromagnetic nature of visible light.

6.2 The equation of continuity

All available experimental evidence suggests that net electric charge cannot be created or destroyed. This *law of conservation of electric charge* can be expressed mathematically as the *equation of continuity*: suppose that a closed surface S encloses volume V. Because electric charge cannot be created or destroyed, the rate of passage (i.e. flux) of electric charge through S is equal to the rate at which electric charge vanishes from V. The flux through S is given in section 3.6, so

$$\oint \boldsymbol{J} \cdot \mathrm{d}\boldsymbol{S} = -\frac{\mathrm{d}}{\mathrm{d}t}\int_V \varrho(\boldsymbol{r})\,\mathrm{d}\tau. \tag{6.5}$$

Because the surface does not vary with time, the orders of integration and differentiation on the right-hand side of eq. (6.5) can be reversed,

$$\oint \boldsymbol{J} \cdot \mathrm{d}\boldsymbol{S} = -\int_V \frac{\partial \varrho(\boldsymbol{r})}{\partial t}\,\mathrm{d}\tau$$

and using the divergence theorem of Appendix A to relate the surface integral on the left-hand side to a certain volume integral we find

$$\int_V \boldsymbol{\nabla} \cdot \boldsymbol{J}\,\mathrm{d}\tau = -\int_V \frac{\partial \varrho(\boldsymbol{r})}{\partial t}\,\mathrm{d}\tau.$$

No assumption has been made about the surface S or the volume V, so the integrands must be equal:

$$\boldsymbol{\nabla} \cdot \boldsymbol{J} = -\frac{\partial \varrho(\boldsymbol{r})}{\partial t}. \tag{6.6}$$

Equation (6.6) is called the *equation of continuity*. It provides a mathematical statement of the law of conservation of electric charge. A similar equation appears in many branches of science and engineering where flow is involved: for example, the corresponding equation in fluid dynamics expresses the physical fact that mass is conserved.

For steady conditions, eq. (6.6) becomes $\boldsymbol{\nabla} \cdot \boldsymbol{J} = 0$.

6.3 The displacement current

Maxwell realized that the equation of continuity and Ampère's law $\boldsymbol{\nabla} \times \boldsymbol{B} = \mu_0 \boldsymbol{J}$ were not compatible in the case where the fields varied with time. Taking the divergence of each side of eq. (6.4) gives

$$\boldsymbol{\nabla} \cdot (\boldsymbol{\nabla} \times \boldsymbol{B}) = \mu_0 \boldsymbol{\nabla} \cdot \boldsymbol{J}$$

the left-hand side of which is zero by a vector identity. Thus in situations where the rate of change of charge density is not zero, Ampère's law contradicts the equation of continuity. Maxwell assumed that Ampère's law should be rewritten

$$\boldsymbol{\nabla} \times \boldsymbol{B} = \mu_0 \boldsymbol{J} + \boldsymbol{X}$$

where the vector $\boldsymbol{X}$ is as yet unknown. The identity of $\boldsymbol{X}$ can be *inferred* by various arguments, but not *proved*. One argument suggesting the identity of

X is as follows. Gauss' theorem expressed in terms of the electric displacement $\boldsymbol{D}$ is

$$\nabla \cdot \boldsymbol{D} = \varrho_{\mathrm{f}}$$

which can be combined with eq. (6.6) to give

$$-\nabla \cdot \boldsymbol{J} = \frac{\partial}{\partial t}(\nabla \cdot \boldsymbol{D})$$

whence

$$\nabla \cdot \left(\boldsymbol{J} + \frac{\partial \boldsymbol{D}}{\partial t} \right) = 0. \tag{6.7}$$

This strongly *suggests* (and nothing more) that $\boldsymbol{X} = \partial \boldsymbol{D}/\partial t$, for $\boldsymbol{J} + \partial \boldsymbol{D}/\partial t$ is a vector field whose divergence is always zero.

Some care is needed, however, because we have used an electro*static* relationship

$$\nabla \cdot \boldsymbol{D} = \varrho_{\mathrm{f}}$$

to derive a *non-static* result.

Whist there are many corresponding plausible arguments that suggest the truth of

$$\nabla \times \boldsymbol{B} = \mu_0 \left[\boldsymbol{J} + \frac{\partial \boldsymbol{D}}{\partial t} \right] \tag{6.8}$$

there is no absolute proof of its validity. All deductions made by assuming the truth of eq. (6.8), however, agree with experimental findings.

Thus, such an alternative line of argument is based on the circuit of Fig. 6.1.

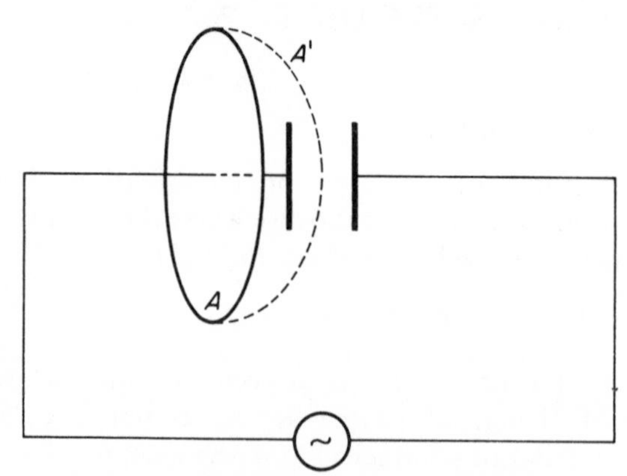

Fig. 6.1 A possible circuit to suggest the existence of the displacement current. An alternating current is applied to the plates of the capacitor and Ampère's law is applied to the surfaces A, A'

An alternating current passes through the capacitor shown, although no charge is transferred between the plates. If the current in the circuit is I, then

$$\oint \boldsymbol{B} \cdot \mathrm{d}\boldsymbol{l} = \mu_0 I$$

where the line integral is over *any* closed path linking the circuit. So Ampère's circuital law, eq. (3.13), takes the particular form

$$\oint_{\mathrm{A}} \boldsymbol{B} \cdot \mathrm{d}\boldsymbol{l} = \mu_0 I$$

for the closed path labelled in A in Fig. 6.1. But at first sight

$$\oint_{\mathrm{A}'} \boldsymbol{B} \cdot \mathrm{d}\boldsymbol{l} = 0$$

since no current flows through the capacitor. This contradiction can only be remedied if an *extra* current is present: for a parallel-plate capacitor

$$E = \sigma/\epsilon_0; \ \sigma = \epsilon_0 E$$

and the change in charge density on the plates is a current density, so

$$\frac{\partial \sigma}{\partial t} = \epsilon_0 \frac{\partial E}{\partial t} = \frac{\partial D}{\partial t}$$

suggesting that the extra current present, even *in vacuo* when no charges move in the space between the plates, is

$$\frac{\partial \boldsymbol{D}}{\partial t}.$$

$\boldsymbol{J}$ is called the *conduction current*, because $\boldsymbol{J} = \varkappa \boldsymbol{E}$ for a conductor which obeys Ohm's law where $\varkappa$ is the *electrical conductivity* (which may well, in the general case be a tensor property). The new current $\partial \boldsymbol{D}/\partial t$ is called the *displacement current.*

Displacement currents are not usually as important as the currents arising from the motion of free charges, but if $\boldsymbol{J} = \boldsymbol{0}$ then the displacement current is the only current present. As we shall see in Section 6.4 it is the existence of the displacement current that permits the propagation of electromagnetic radiation. A summary of Maxwell's equations in terms of the fields $\boldsymbol{E}$ and $\boldsymbol{B}$ can be presented either as differential equations

$$\left.\begin{array}{ll} \nabla \cdot \boldsymbol{E} = \varrho/\epsilon_0 & \nabla \times \boldsymbol{E} = -\dfrac{\partial \boldsymbol{B}}{\partial t} \\ \nabla \cdot \boldsymbol{B} = 0 & \nabla \times \boldsymbol{B} = \mu_0 \left[\boldsymbol{J} + \dfrac{\partial \boldsymbol{D}}{\partial t}\right] \end{array}\right\} \tag{6.9}$$

or in their entirely equivalent integral forms

$$\left.\begin{aligned} &\oint \boldsymbol{E}\cdot \mathrm{d}\boldsymbol{S} = Q/\epsilon_0 \qquad && \int \boldsymbol{E}\cdot \mathrm{d}\boldsymbol{l} = -\frac{\partial}{\partial t}\int \boldsymbol{B}\cdot \mathrm{d}\boldsymbol{S} \\ &\oint \boldsymbol{B}\cdot \mathrm{d}\boldsymbol{S} = 0 && \int \boldsymbol{B}\cdot \mathrm{d}\boldsymbol{l} = \mu_0 \int \left[\boldsymbol{J} + \frac{\partial \boldsymbol{D}}{\partial t}\right]\cdot \mathrm{d}\boldsymbol{S} \end{aligned}\right\} \qquad (6.10)$$

We will in general work with the differential forms (6.9). It is, however, important to understand that Maxwell's equations, irrespective of the way they are written, do not yield $\boldsymbol{E}$ and $\boldsymbol{B}$ directly: the sets of eqs. (6.9) or (6.10) have to be solved with the appropriate boundary conditions. Also needed are the constitutive relations such as

$$\boldsymbol{P} = \epsilon_0 \chi_e \boldsymbol{E}.$$

As formulated, the equations are valid for media at rest with respect to the observer: for moving media the equation

$$\nabla \times \boldsymbol{B} = \mu_0 \left[\boldsymbol{J} + \frac{\partial \boldsymbol{D}}{\partial t}\right]$$

needs some reinterpretation although, unlike the equations of classical newtonian mechanics, Maxwell's equations *do* satisfy the requirements of the special theory of relativity (Chapter 7).

Maxwell's equations apply to *all* electromagnetic phenomena in *all* materials. Following the arguments advanced in Chapters 2 and 4, when dealing with dielectric and magnetic materials, the differential forms of Maxwell's equations (6.9) are often usefully rewritten so that only the *free* charges and *free* currents, ϱ_f and $\boldsymbol{J}_f$ introduced in those chapters, appear in the equations. It is left as an exercise to verify that the corresponding expressions are

$$\left.\begin{aligned} &\nabla\cdot \boldsymbol{D} = \varrho_f \qquad && \nabla \times \boldsymbol{E} = -\frac{\partial \boldsymbol{B}}{\partial t} \\ &\nabla\cdot \boldsymbol{B} = 0 && \nabla \times \boldsymbol{H} = \boldsymbol{J}_f + \frac{\partial \boldsymbol{D}}{\partial t} \end{aligned}\right\} \qquad (6.11)$$

The sets of eqs. (6.9) and (6.11) are exactly correct; no assumptions have been made whatever about the constitutive relations. We have, however, assumed in the spirit of classical electromagnetism that any differential volumes are sufficiently large or properly averaged so as to be homogeneous: in other words we have ignored the fine details of molecular structure.

Broadly speaking, molecular charge distributions need to be calculated according to the laws of quantum mechanics. Once the molecular charge distribution is known, Maxwell's equations apply. The *molecular* fields $\boldsymbol{E}_m$ and

$\boldsymbol{B}_{\rm m}$ obey a set of equations completely analogous to the set (6.9):

$$\nabla\cdot\boldsymbol{E}_{\rm m}=\varrho_{\rm m}/\epsilon_0 \qquad \nabla\times\boldsymbol{E}_{\rm m}=-\frac{\partial\boldsymbol{B}_{\rm m}}{\partial t}$$

$$\nabla\cdot\boldsymbol{B}_{\rm m}=0 \qquad \nabla\times\boldsymbol{B}_{\rm m}=\mu_0\left[\boldsymbol{J}_{\rm m}+\frac{\partial\boldsymbol{D}_{\rm m}}{\partial t}\right]$$

and the 'macroscopic' differential equations (6.9) are derived from this set by averaging over volumes that are very small in the absolute sense, but which contain very many molecules.

At this point it is worthwhile to recall the physical content of Maxwell's equations.

The inverse square law of electrostatics leads to Gauss' theorem $\nabla\cdot\boldsymbol{E}=\varrho/\epsilon_0$: if dielectrics are present they acquire a polarization $\boldsymbol{P}$ and are equivalent for the purposes of calculating effects outside the dielectric sample to a volume charge density $-\nabla\cdot\boldsymbol{P}$. In order that Gauss' theorem may be written in terms of the free charges, one defines the electric displacement $\boldsymbol{D}=\epsilon_0\boldsymbol{E}+\boldsymbol{P}$. Hence Gauss' theorem becomes

$$\nabla\cdot\boldsymbol{D}=\varrho_{\rm free}$$

The equation $\nabla\cdot\boldsymbol{B}=0$ or its equivalent integral form $\int\boldsymbol{B}\cdot{\rm d}\boldsymbol{S}=0$ has the physical interpretation that there are no 'sources' of $\boldsymbol{B}$, i.e. magnetic monopoles do not exist. $\nabla\times\boldsymbol{E}=-(\partial\boldsymbol{B}/\partial t)$ simply says that a changing magnetic induction yields an electric field.

Ampère's law $\nabla\times\boldsymbol{H}=\boldsymbol{J}_{\rm f}$ for steady currents summarizes the results of experiments which showed that small coils of wire carrying steady currents exert forces on each other which show the same behaviour as two permanent magnets. Magnetic materials when present acquire a magnetization $\boldsymbol{M}$, and the magnetic field $\boldsymbol{H}$ is introduced in order that Ampère's law may be written in terms of the free currents. If the currents vary in time, the displacement current has to be added giving finally

$$\nabla\times\boldsymbol{H}=\boldsymbol{J}_{\rm f}+\frac{\partial\boldsymbol{D}}{\partial t}.$$

6.4 Electromagnetic waves in free space

In this section we demonstrate the existence of solutions of Maxwell's equations in free space that correspond to electromagnetic waves travelling at the speed of light. In free space, with no current or charge densities present, eqs (6.9) take the form

$$\nabla\cdot\boldsymbol{E}=0 \qquad (6.12)$$

$$\nabla\cdot\boldsymbol{B}=0 \qquad (6.13)$$

$$\nabla \times \boldsymbol{E} = -\frac{\partial \boldsymbol{B}}{\partial t} \tag{6.14}$$

$$\nabla \times \boldsymbol{B} = \epsilon_0 \mu_0 \frac{\partial \boldsymbol{E}}{\partial t} \tag{6.15}$$

Taking the curl of eqs (6.14) and (6.15) gives

$$\nabla \times (\nabla \times \boldsymbol{E}) = -\nabla \times \frac{\partial \boldsymbol{B}}{\partial t} \tag{6.16}$$

$$\nabla \times (\nabla \times \boldsymbol{B}) = \epsilon_0 \mu_0 \nabla \times \frac{\partial \boldsymbol{E}}{\partial t}. \tag{6.17}$$

Interchanging the order of differentiation with respect to space and time in eqs (6.16) and (6.17) gives

$$\nabla \times (\nabla \times \boldsymbol{E}) = -\frac{\partial}{\partial t} (\nabla \times \boldsymbol{B}) \tag{6.18}$$

$$\nabla \times (\nabla \times \boldsymbol{B}) = -\epsilon_0 \mu_0 \frac{\partial}{\partial t} (\nabla \times \boldsymbol{E}) \tag{6.19}$$

Substituting (6.14) into the right-hand side of eq. (6.18) and likewise eq. (6.15) into eq. (6.17), and using the vector identity

$$\text{curl curl} \equiv \text{grad div} - \nabla^2$$

$$\nabla (\nabla \cdot \boldsymbol{B}) - \nabla^2 \boldsymbol{B} = -\epsilon_0 \mu_0 \frac{\partial^2 \boldsymbol{B}}{\partial t^2}$$

$$\nabla (\nabla \cdot \boldsymbol{E}) - \nabla^2 \boldsymbol{E} = -\epsilon_0 \mu_0 \frac{\partial^2 \boldsymbol{E}}{\partial t^2}.$$

From eqs (6.12) and (6.13), $\nabla \cdot \boldsymbol{E} = \nabla \cdot \boldsymbol{B} = 0$, hence

$$\nabla^2 \boldsymbol{B} = \epsilon_0 \mu_0 \frac{\partial^2 \boldsymbol{B}}{\partial t^2} \tag{6.20}$$

$$\nabla^2 \boldsymbol{E} = \epsilon_0 \mu_0 \frac{\partial^2 \boldsymbol{E}}{\partial t^2} \tag{6.21}$$

which are the wave equations for $\boldsymbol{E}$ and $\boldsymbol{B}$.

The speed of propagation of these waves is $(\epsilon_0 \mu_0)^{-1/2}$: the constant μ_0 is *defined* to be $4\pi \times 10^{-7}$ H m^{-1} and the permittivity of free space ϵ_0 is found from electrostatic measurements to be (about) 8.854×10^{-12} F m^{-1}, giving a wave speed of 2.998×10^8 m s^{-1}, the speed of light *in vacuo*. Moreover, according to Maxwell's equations, *all* electromagnetic waves propagate in free space at the speed of light c_0. We note that, if the displacement current were *not* present, no electromagnetic wave solutions of Maxwell's equations would exist because eq. (6.15) would give $\nabla \times \boldsymbol{B} = \boldsymbol{0}$.

From now on in this chapter we follow the custom of using $\boldsymbol{H}$ rather than $\boldsymbol{B}$ when discussing electromagnetic waves: a reason is that it turns out that vector $\boldsymbol{E}\times\boldsymbol{H}$ has a special significance in the study of energy propagation rather than $\boldsymbol{E}\times\boldsymbol{B}$. We now study a very simple solution of the wave eqs (6.20) and (6.21): the *plane wave* (to be contrasted with the *attenuated wave* of Section 6.7). A plane wave is one whose amplitude maxima are constant over all points of a plane drawn normal to the direction of propagation. We first show that such waves consist of transverse vibrations of electric and magnetic fields, which can vibrate at right angles to the direction of propagation: consider a monochromatic plane wave moving along the z-axis with electric vector given by

$$\boldsymbol{E}=\boldsymbol{E}_0 \exp i(kz-\omega t), \tag{6.22}$$

where the wavenumber $k=\omega/c$ with ω the radian frequency $2\pi\nu$, and $\boldsymbol{E}_0$ an arbitrary constant amplitude vector. We will see later that the choice of z-axis as the direction of propagation involves no loss of generality. We first demonstrate that $\boldsymbol{E}_0$ must lie in a plane parallel to the (xz) plane. Equation (6.12) gives $\nabla\cdot\boldsymbol{E}=0$, and because $\boldsymbol{E}_0$ is a vector with constant modulus

$$\frac{\partial E_x}{\partial x}=\frac{\partial E_y}{\partial y}=0.$$

Thus $\partial E_z/\partial z=0$ which means that the z component of $\boldsymbol{E}$ is independent of z. To reconcile this with eq. (6.22), $E_z=0$ and hence $\boldsymbol{E}_0$ must be perpendicular to the direction of propagation. A similar argument applies to the magnetic field $\boldsymbol{H}$: if

$$\boldsymbol{H}=\boldsymbol{H}_0 \exp i\,(kz-\omega t+\phi) \tag{6.23}$$

where ϕ is the phase (as yet undetermined) with respect to $\boldsymbol{E}$, $\boldsymbol{H}_0$ is also perpendicular to the direction of propagation. To simplify the argument (without loss of generality), assume that $\boldsymbol{E}_0$ lies in the (xz) plane. Thus

$$\boldsymbol{E}=E_0 \exp i\,(kz-\omega t)\,\hat{\boldsymbol{x}}$$

where $\hat{\boldsymbol{x}}$ is a unit vector in the $+x$ direction. We now show that $\boldsymbol{E}$ and $\boldsymbol{H}$ are in phase. We have

$$\nabla\times\boldsymbol{E}=-\frac{\partial\boldsymbol{B}}{\partial t}$$

so

$$-\frac{\partial\boldsymbol{B}}{\partial t}=ik \exp i\,(kz-\omega t)\,\hat{\boldsymbol{y}}$$

and on integrating with respect to time we find

$$\boldsymbol{B}=\frac{k}{\omega}\,E_0 \exp i\,(kz-\omega t)\,\hat{\boldsymbol{y}}$$

or, in terms of $\boldsymbol{H}$

$$\boldsymbol{H} = \frac{\omega}{k} \epsilon_0 E_0 \exp i (kz - \omega t) \hat{\boldsymbol{y}}. \tag{6.24}$$

This demonstrates the important conclusion that the electric and magnetic fields oscillate in phase. Equation (6.24) can be written

$$\boldsymbol{H} = \left(\frac{\epsilon_0}{\mu_0}\right)^{1/2} E\hat{\boldsymbol{y}} \tag{6.25}$$

and so

$$\boldsymbol{E} \times \boldsymbol{H} = \left(\frac{\epsilon_0}{\mu_0}\right)^{1/2} \hat{\boldsymbol{z}} \tag{6.26}$$

showing that $\boldsymbol{E}$ and $\boldsymbol{H}$ are perpendicular and also that $\boldsymbol{E} \times \boldsymbol{H}$ points along the axis of propagation of the wave. The ratio of the amplitudes of the electric and magnetic fields is $(\epsilon_0/\mu_0)^{1/2}$. Equations (6.25) and (6.26) can be written in terms of $\boldsymbol{E}$ and $\boldsymbol{B}$

$$\boldsymbol{B} = \frac{1}{c} E\hat{\boldsymbol{y}}$$

$$\boldsymbol{E} \times \boldsymbol{B} = \frac{1}{c} \hat{\boldsymbol{z}}$$

showing that the ratio of the electric field to the magnetic induction is c, the speed of light.

In the general case where the radiation propagates along a direction defined by unit vector $\hat{\boldsymbol{k}}$, the eqs (6.25) and (6.26) appropriate to monochromatic radiation are obtained from

$$\boldsymbol{E} = \boldsymbol{E}_0 \exp i(\boldsymbol{k} \cdot \boldsymbol{r} - \omega t) \tag{6.27}$$

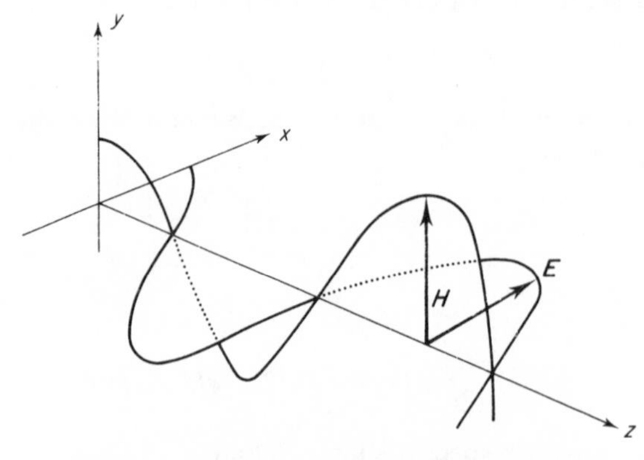

Fig. 6.2 Electric and magnetic vectors associated with electromagnetic radiation. The fields are self-propagating

as

$$H = \left(\frac{\epsilon_0}{\mu_0}\right)^{1/2} \hat{k} \times E \tag{6.28}$$

$$E \times H = \left(\frac{\epsilon_0}{\mu_0}\right)^{1/2} \hat{k}. \tag{6.29}$$

6.5 Polarized radiation

So far we have restricted our discussion of the solution of the wave equations (6.20) and (6.21) to the simple case of plane polarized radiation. In fact this is sufficient for the general so! 'tion because the wave equations are *linear* in E and H: so for example, i: E_1 nd E_2 are two separate solutions of eq. (6.21), i.e.

$$\nabla^2 E_1 = \frac{1}{c^2}\frac{\partial^2 E_1}{\partial t^2}$$

$$\nabla^2 E_2 = \frac{1}{c^2}\frac{\partial^2 E_2}{\partial t^2}$$

then since

$$\nabla^2(a_1E_1 + a_2E_2) = \frac{1}{c^2}\frac{\partial^2}{\partial t^2}(a_1E_1 + a_2E_2)$$

any linear combination of E_1 and E_2 is also a solution of eq. (6.21). Naturally the same comment holds for the H field. Thus a monochromatic electromagnetic wave could be formed from a superposition of very many different components having different amplitudes $E_{0,j}$ and phases ϕ_j:

$$E = \sum_j E_{0,j} \exp i(k \cdot r - \omega t + \phi_j) \tag{6.30}$$

and the composition of E (or H) is referred to as the *state of polarization* of the radiation. We have already met the equation

$$E = \hat{x}E_0 \exp i(kz - \omega t)$$

for a monochromatic wave propagating along the z direction with its electric field oscillating in the (xz) plane. Such radiation is often called *linearly polarized radiation*. A linearly polarized monochromatic wave whose electric field is in an arbitrary plane containing the z-axis can be resolved in two components

$$E = (\hat{x}E_{0,x} + \hat{y}E_{0,y}) \exp i(kz - \omega t) \tag{6.31}$$

corresponding to two waves in phase and polarized at 90° to each other. The electric field

$$E = \hat{x}E_{0,x} \exp i(kz - \omega t) + \hat{y}E_{0,y} \exp i(kz - \omega t + \pi/2) \tag{6.32}$$

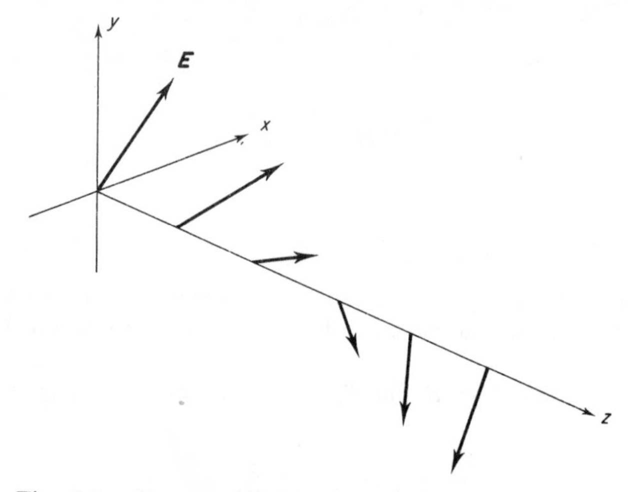

Fig. 6.3 Circularly polarized radiation. The electric (and magnetic) vectors describe a circle about the z-axis

represents *ellipitically polarized radiation*: the direction of $\boldsymbol{E}$ changes as the wave propagates along the z-axis in such a way as to describe an ellipse. In the special case $E_{0,x} = E_{0,y}$ the $\boldsymbol{E}$ vector describes a circle as the wave propagates, and the radiation is called *circularly polarized radiation*.

A *randomly polarized*, or *unpolarized* wave is one in which the direction of $\boldsymbol{E}$ is changing randomly with time.

An experimental use for polarized radiation is in investigating directional effects: if a substance under study is not isotropic it will interact differently with the radiation as the plane of polarization is changed, hence giving information about anisotropy (cf p. 206).

6.6 Electromagnetic waves in matter

In the presence of matter the electromagnetic wave problem can become very much more difficult. We first treat linear homogeneous isotropic media (a medium is *homogeneous* if its properties do not vary from point to point: it is *isotropic* if its properties are the same in all directions, and the adjective *linear* implies linear response of $\boldsymbol{M}$ and $\boldsymbol{P}$ to applied fields). We also restrict the discussion to media that are non-conducting, so that $\boldsymbol{J}_f = \boldsymbol{0}$. Starting from Maxwell's eqs (6.9) and repeating the arguments of Section 6.4 we find, on substituting the constitutive relations

$$\boldsymbol{D} = \epsilon \boldsymbol{E}; \qquad \boldsymbol{H} = \boldsymbol{B}/\mu$$

that

$$\nabla^2 \boldsymbol{E} - \mu\epsilon \frac{\partial^2 \boldsymbol{E}}{\partial t^2} - \boldsymbol{\nabla}\left(\frac{\varrho_f}{\epsilon}\right) = 0 \qquad (6.33)$$

$$\nabla^2 H - \mu\epsilon \frac{\partial^2 H}{\partial t^2} = 0 \qquad (6.34)$$

The eqs (6.33) and (6.34) are similar: eq. (6.34) lacks a term equivalent to ϱ_f because there is no magnetic equivalent of electric charge. We again consider for simplicity the case of a plane wave propagating along the z-axis. Then

$$\nabla \cdot E = \frac{\partial E_z}{\partial z} = \varrho_f/\epsilon_0$$

Thus

$$\nabla(\varrho_f/\epsilon) = \nabla\left(\frac{\partial E_z}{\partial z}\right) = \hat{z}\frac{\partial^2 E_z}{\partial z^2}.$$

The wave eq. (6.33) for E then becomes

$$\nabla^2 E - \mu\epsilon \frac{\partial^2 E}{\partial t^2} - \frac{\partial^2 E_z}{\partial z^2}\hat{z} = 0.$$

The longitudinal component E_z of this equation must therefore satisfy

$$\frac{\partial^2 E_z}{\partial z^2} - \mu\epsilon \frac{\partial^2 E_z}{\partial t^2} - \frac{\partial^2 E_z}{\partial z^2} = 0$$

i.e.

$$\mu\epsilon \frac{\partial^2 E_z}{\partial t^2} = 0. \qquad (6.35)$$

Hence $E_z = a + bt$, where a and b are constants of integration. This does not describe a wave solution unless $a = b = 0$, so $E_z = 0$. The H vector can also be shown to be transverse, and in fact *plane waves are transverse in any homogeneous, linear isotropic non-conducting medium*. By repeating the argument of Section 6.5 it is possible to show also that the E and H vectors are mutually perpendicular and oriented so that $E \times H$ points in the direction of propagation. Also, on substituting eq. (6.35) into eq. (6.33) we recognize that the speed of such waves is

$$\left(\frac{1}{\epsilon\mu}\right)^{1/2} = \frac{c}{(\epsilon_r\mu_r)^{1/2}}. \qquad (6.36)$$

In free space where $\epsilon_r = \mu_r = 1$, this velocity increases to c, the speed of light *in vacuo*. In a non-magnetic medium $\mu_r = 1$, and the ratio of this velocity to the speed of light *in vacuo* is called the *refractive index n* of the medium. Thus

$$n = (\epsilon_r)^{1/2}.$$

However, it turns out that, for real substances, ϵ_r and hence n varies with the frequency of the radiation used. This feature has not so far appeared in our treatment because we have not taken sufficiently detailed account of the molecular structure of the dielectric. A full discussion is deferred to Chapter 9.

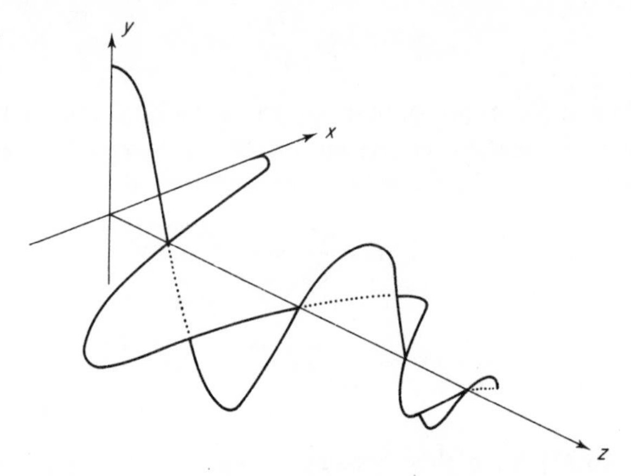

Fig. 6.4 Attenuation of a plane polarized wave

The discussion above has clearly treated idealized limiting cases. The process of solving Maxwell's equations when real media are present is a very complicated problem. We close the discussion, however, by indicating the corresponding solution of Maxwell's equations for an isotropic, homogeneous linear *conducting* medium, because an important new feature, *attenuation*, appears. For a conductor obeying Ohm's law, $\boldsymbol{J} = \varkappa \boldsymbol{E}$, Maxwell's equations become

$$\nabla^2 \boldsymbol{E} - \mu\varkappa \frac{\partial \boldsymbol{E}}{\partial t} - \mu\epsilon \frac{\partial^2 \boldsymbol{E}}{\partial t^2} = 0$$

$$\nabla^2 \boldsymbol{H} - \mu\varkappa \frac{\partial \boldsymbol{H}}{\partial t} - \mu\epsilon \frac{\partial^2 \boldsymbol{H}}{\partial t^2} = 0.$$

The detailed solution of these equations is not our concern: the electric and magnetic fields still have the main features of the waves discussed above, but the electric vector, for example, varies as follows

$$\boldsymbol{E} = \boldsymbol{E}_0 \exp i(kz - \omega t) \exp(-\beta z),$$

where $\beta = \sqrt{(\omega\mu\varkappa/2)} > 0$. The maximum amplitude of the electric (and magnetic) vector thus decreases with travel along the z-axis, and this process is called *attenuation*. A wave cannot propagate in a conducting medium without attenuation.

6.7 Propagation of energy by electromagnetic waves

In previous sections we saw that the energy density associated with a static electric field is

$$w_{\text{el}} = \tfrac{1}{2} \boldsymbol{D} \cdot \boldsymbol{E} \tag{6.37}$$

and that the energy density associated with a magnetostatic field was

$$w_{\mathrm{mag}} = \tfrac{1}{2}\boldsymbol{B}\cdot\boldsymbol{H}. \tag{6.38}$$

These expressions are quite general since no assumptions were made about the constitutive relations in their deduction. The total energy U in a combination of electrostatic and magnetostatic fields is thus

$$U = \frac{1}{2}\int_V (\boldsymbol{D}\cdot\boldsymbol{E} + \boldsymbol{B}\cdot\boldsymbol{H})\,\mathrm{d}\tau, \tag{6.39}$$

where V is the volume over which the fields exist. Equation (6.39) also applies when the fields vary with time.

Since electromagnetic waves consist of varying electric and magnetic fields, we therefore know the energy associated with the fields. We also know that the waves propagate along the direction of $\boldsymbol{E}\times\boldsymbol{H}$, so we now investigate whether the vector $\boldsymbol{E}\times\boldsymbol{H}$ has any physical connection with energy.

Let us calculate the flux of $\boldsymbol{E}\times\boldsymbol{H}$ through the arbitrary closed surface S:

$$\begin{aligned} \text{Flux} &= \oint_S (\boldsymbol{E}\times\boldsymbol{H})\cdot\mathrm{d}\boldsymbol{S} \\ &= \int_V \nabla\cdot(\boldsymbol{E}\times\boldsymbol{H})\,\mathrm{d}\tau \end{aligned} \tag{6.40}$$

where the divergence theorem has been used to express the surface integral in terms of a volume integral, the surface S containing the volume V.

Since

$$\nabla\cdot(\boldsymbol{E}\times\boldsymbol{H}) = \boldsymbol{H}\cdot(\nabla\times\boldsymbol{E}) - \boldsymbol{E}\cdot(\nabla\times\boldsymbol{H})$$

by a certain vector identity,

$$\int_S (\boldsymbol{E}\times\boldsymbol{H})\cdot\mathrm{d}\boldsymbol{S} = \int [\boldsymbol{H}\cdot(\nabla\times\boldsymbol{E}) - \boldsymbol{E}\cdot(\nabla\times\boldsymbol{H})]\,\mathrm{d}\tau$$

and since

$$\nabla\times\boldsymbol{E} = -\frac{\partial\boldsymbol{B}}{\partial t}; \qquad \nabla\times\boldsymbol{H} = \boldsymbol{J}_{\mathrm{f}} + \partial\boldsymbol{D}/\partial t$$

where in free space $\boldsymbol{J}_{\mathrm{f}} = \boldsymbol{0}$

$$\int_S (\boldsymbol{E}\times\boldsymbol{H})\cdot\mathrm{d}\boldsymbol{S} = -\frac{\partial}{\partial t}\int_V [\boldsymbol{H}\cdot\boldsymbol{B} + \boldsymbol{E}\cdot\boldsymbol{D}]\,\mathrm{d}\tau. \tag{6.41}$$

This shows that the flux of $\boldsymbol{E}\times\boldsymbol{H}$ through the surface S is the rate at which energy is lost from the volume V. The quantity

$$\boldsymbol{S} = \boldsymbol{E}\times\boldsymbol{H} \tag{6.42}$$

is called the *Poynting vector*: when integrated over a closed surface it gives the total rate of outward flow of energy. In the case of sinusoidally varying fields the average value of $\boldsymbol{S}$ is

$$\langle S \rangle = \tfrac{1}{2} \boldsymbol{E}_0 \times \boldsymbol{H}_0$$

the factor of $\frac{1}{2}$ appearing because the average value of sin x over one oscillation is $(\frac{1}{2})^{-1/2}$. In Section 6.10 we discuss the principles involved in the propagation and transmission of electromagnetic radiation.

6.8 The wave equations for the potentials

As noted in Chapters 1 and 3, it is often more fruitful to recast the basic electromagnetic field equations in terms of the scalar and the vector potentials V and $\boldsymbol{A}$, where

$$\left.\begin{aligned} \boldsymbol{B} &= \nabla \times \boldsymbol{A} \\ \boldsymbol{E} &= -\nabla V - \frac{\partial \boldsymbol{A}}{\partial t} \end{aligned}\right\} \tag{6.43}$$

Starting from Maxwell's eqs (6.8) written in terms of the fields $\boldsymbol{E}$ and $\boldsymbol{B}$ and substituting eqs (6.43) into Gauss' law yields

$$\nabla \cdot \left(-\nabla V - \frac{\partial \boldsymbol{A}}{\partial t} \right) = \varrho/\epsilon_0;$$

hence

$$-\nabla^2 V - \frac{\partial}{\partial t}(\nabla \cdot \boldsymbol{A}) = \varrho/\epsilon_0. \tag{6.44}$$

On substituting eqs (6.43) into the equation

$$\nabla \times \boldsymbol{B} = \mu_0 \left(\boldsymbol{J} + \epsilon \frac{\partial \boldsymbol{E}}{\partial t} \right)$$

we find

$$\nabla \times (\nabla \times \boldsymbol{A}) = \mu_0 \boldsymbol{J} - \mu_0 \epsilon_0 \left(\frac{\partial^2 \boldsymbol{A}}{\partial t^2} + \nabla \frac{\partial V}{\partial t} \right)$$

or

$$\nabla (\nabla \cdot \boldsymbol{A}) - \nabla^2 \boldsymbol{A} = \mu_0 \boldsymbol{J} - \frac{1}{c^2} \frac{\partial^2 \boldsymbol{A}}{\partial t^2} - \frac{1}{c^2} \nabla \frac{\partial V}{\partial t}. \tag{6.45}$$

As mentioned earlier, the vector potential $\boldsymbol{A}$ is not uniquely defined by $\boldsymbol{B} = \nabla \times \boldsymbol{A}$. We therefore make a convenient choice of gauge by requiring

$$\nabla \cdot \boldsymbol{A} = -\frac{1}{c^2} \frac{\partial V}{\partial t}; \tag{6.46}$$

this is called the *Lorentz gauge*. Hence eq. (6.44) and eq. (6.45) become

$$\nabla^2 V = \frac{1}{c^2}\frac{\partial^2 V}{\partial t^2} = -\varrho/\epsilon_0 \tag{6.47}$$

$$\nabla^2 \boldsymbol{A} - \frac{1}{c^2}\frac{\partial^2 \boldsymbol{A}}{\partial t^2} = -\mu_0 \boldsymbol{J} \tag{6.48}$$

and in free space where $\varrho = 0$ and $\boldsymbol{J} = \boldsymbol{0}$ we obtain wave equations very similar to those already encountered.

Equation (6.47) and (6.48) are often written more compactly in terms of the d'Alembertian operator

$$\Box^2 \equiv \nabla^2 - \frac{1}{c^2}\frac{\partial^2}{\partial t^2}$$

so that

$$\Box^2 \boldsymbol{A} = -\mu_0 \boldsymbol{J}; \qquad \Box^2 V = -\varrho/\epsilon_0 \tag{6.49}$$

which equations are very similar to Poisson's equation, eq. (1.38).

6.9 The retarded potentials

The wave equations (6.47) and (6.48) for $\boldsymbol{A}$ and V have a serious flaw: they do not take into account the finite velocity of propagation of electric and magnetic fields: as we shall see in Chapter 7, such fields propagate through free space at the speed of light. We have already found in Chapters 1 and 3 solutions for the potentials in the static case:

$$V(\boldsymbol{r}) = \frac{1}{4\pi\epsilon_0}\int_{V'} \frac{\varrho(\boldsymbol{r}')\,\mathrm{d}\tau'}{|\boldsymbol{r} - \boldsymbol{r}'|} \tag{6.50}$$

$$\boldsymbol{A}(\boldsymbol{r}) = \frac{\mu_0}{4\pi}\int_{V'} \frac{\boldsymbol{J}(\boldsymbol{r}')\,\mathrm{d}\tau'}{|\boldsymbol{r} - \boldsymbol{r}'|} \tag{6.51}$$

so the *general* expressions must reduce to eqs (6.50) and (6.51) when all time derivatives are zero. Consider the case where the electric charge density varies with time; because fields propagate at the speed of light, the change in ϱ will only manifest itself at the field point $\boldsymbol{r}$ at a time $|\boldsymbol{r} - \boldsymbol{r}'|/c$ *after* the change, this being the time for the electromagnetic field to travel a distance $|\boldsymbol{r} - \boldsymbol{r}'|$. Hence the contribution to the potential $V(\boldsymbol{r})$ at time t depends on the charge density as it was at time $t - |\boldsymbol{r} - \boldsymbol{r}'|/c$ and so eq. (6.50) needs to be modified:

$$V(\boldsymbol{r}, t) = \frac{1}{4\pi\epsilon_0}\int \frac{\varrho(\boldsymbol{r}', t - |\boldsymbol{r} - \boldsymbol{r}'|/c)}{|\boldsymbol{r} - \boldsymbol{r}'|}\,\mathrm{d}\tau' \tag{6.52}$$

with a corresponding expression for $\boldsymbol{A}(\boldsymbol{r}, t)$. The potentials $V(\boldsymbol{r}, t)$ and $\boldsymbol{A}(\boldsymbol{r}, t)$ are known as the *retarded potentials*, and they are often written in a

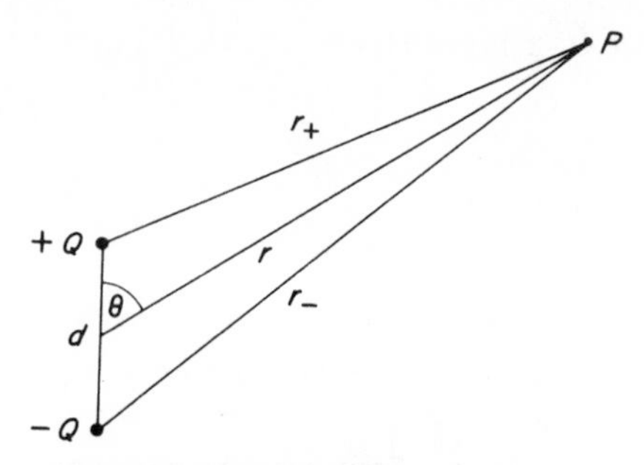

Fig. 6.5 Calculation of the potential and field at point P (taken as origin) due to an oscillating dipole

more compact notation using the retarded values of the variables: thus, for example, the retarded charge density

$$[\varrho(\boldsymbol{r}')] = \varrho(\boldsymbol{r}', t - |\boldsymbol{r} - \boldsymbol{r}'|/c) \tag{6.53}$$

represents the charge density at $\boldsymbol{r}'$ as it was $|\boldsymbol{r} - \boldsymbol{r}'|/c$ ago from the time of the measurement t. Thus for example, eq. (6.52) can be rewritten.

$$V(\boldsymbol{r}, t) = \frac{1}{4\pi\epsilon_0}\int_{V'} \frac{[\varrho(\boldsymbol{r}')]}{|\boldsymbol{r} - \boldsymbol{r}'|}\,\mathrm{d}\tau',$$

where the integral is evaluated over the volume V' as it was $|\boldsymbol{r} - \boldsymbol{r}'|/c$ ago, when it enclosed all the charges. In the general case it is a very difficult matter to give general expressions for the retarded potentials, but in some cases the retardation effects can be negligible. In the following section we illustrate the importance of retardation in the specific case of an oscillating source of radiation and show how the limiting values of the fields can be calculated.

6.10 Dipole radiation

Consider the simple oscillating dipole shown in Fig. 6.5, which consists of two equal and opposite charges a distance d apart, the two charges varying harmonically with time with circular frequency ω. Such a dipole is often referred to as a *hertzian oscillator* and we shall now show that such a dipole radiates electromagnetic waves.

Putting $Q = Q_0 \exp(i\omega t)$ the dipole $\boldsymbol{p}$ is given by

$$\boldsymbol{p} = \boldsymbol{p}_0 \exp(i\omega t) \tag{6.54}$$

with $\boldsymbol{p}_0 = Q_0\boldsymbol{d}$. We first find the retarded potentials for the hertzian dipole, making the assumptions appropriate to a small dipole as in Section 1.6. In this case the field point is taken to be such that $r \gg d$, and we also impose the simplifying condition that the wavelength $\lambda = 2\pi c/\omega$ associated with the oscillation is large compared to the size of the dipole, $\lambda \gg d$. Assuming that the wave propagates in free space, the retarded potential V is

$$V = \frac{Q_0}{4\pi\epsilon_0}\left\{\frac{\exp i\omega(t - r_+/c)}{r_+} - \frac{\exp i\omega(t - r_-/c)}{r_-}\right\} \tag{6.55}$$

and so the charges $\pm Q$ give scalar potentials which differ not only in magnitude but also in phase at the field point. Whilst it *is* possible to expand the exponentials and the denominators in eq. (6.55) in order to find the scalar potential V, the derivation is rather involved because the expression involves the *difference* between two very nearly equal scalar potentials and so great care has to be taken in the cancellation of terms in the expansions. Discussions of dipole radiation usually start by calculating $\boldsymbol{A}$, then $\boldsymbol{B} = \nabla \times \boldsymbol{A}$ and finally finding V from the Lorentz condition

$$\nabla \cdot \boldsymbol{A} + \frac{1}{c^2}\frac{\partial V}{\partial t} = 0$$

and hence $\boldsymbol{E}$ from

$$\boldsymbol{E} = -\frac{\partial \boldsymbol{A}}{\partial t} - \nabla V.$$

When this is done, the expressions for $\boldsymbol{H}$ and $\boldsymbol{E}$ turn out to be, in spherical polar coordinates,

$$\left.\begin{aligned} H_r = 0; \qquad H_\theta = 0 \\ H_\phi = -\frac{cp_0k^2}{4\pi}\sin\theta\left\{1 - \frac{i}{kr}\right\}\exp\left(i\omega[t]\right) \end{aligned}\right\} \tag{6.56}$$

and

$$\left.\begin{aligned} E_\phi &= 0. \\ E_r &= \frac{ip_0k}{2\pi\epsilon_0 r}\cos\theta\left\{1 - \frac{i}{kr}\right\}\exp\left(-i\omega[t]\right) \\ E_\theta &= -\frac{p_0k^2}{4\pi\epsilon_0 r}\sin\theta\left\{1 - \frac{i}{kr}\left(1 - \frac{i}{kr}\right)\right\}\exp\left(-i\omega[t]\right), \end{aligned}\right\} \tag{6.57}$$

where $k = \omega/c$ and $[t] = t - r/c$.

This set of equations is known as *Hertz's relations*. It is very informative to study the limiting behaviour of $\boldsymbol{H}$ and $\boldsymbol{E}$, corresponding to points very near to the oscillating dipole and to points very far away from the dipole. If $r \ll \lambda$ (or $kr \ll 1$) one speaks of the *near zone*, and if $kr \gg 1$ one speaks of the *far zone*, or *radiation zone*.

Near zone

$$E_r \sim \frac{p\cos\theta}{2\pi\epsilon_0 r^3}. \qquad E_\theta \sim \frac{p\sin\theta}{4\pi\epsilon_0 r^3}; \qquad H_\phi \sim \frac{i\omega p\sin\theta}{4\pi r^2} \tag{6.58}$$

showing that the electric field is simply a dipole field emanating from an electric dipole $\boldsymbol{p}$ located at the origin. Although the dipole is an oscillating dipole, the field point is so close to the source that retardation effects are negligible. The expression for $\boldsymbol{H}$ can be rewritten

$$H_\phi \sim \frac{1}{4\pi r^2}\frac{\mathrm{d}\boldsymbol{p}}{\mathrm{d}t}\sin\theta.$$

Radiation zone

The dominant terms are now

$$\left.\begin{aligned} E_\theta &\sim \frac{k^2\sin\theta}{4\pi\epsilon_0 r}\,p_0\exp\,(i\omega[t]) \\ H_\phi &\sim \frac{ck^2\sin\theta}{4\pi r}\,p_0\exp\,(i\omega[t]) \end{aligned}\right\} \tag{6.59}$$

showing that, at large distances from the source, the electric and magnetic fields are transverse and fall off as $1/r$. This $1/r$ dependence is due entirely to retardation, and we shall see later that this dependence leads to conservation of energy.

Considering $\boldsymbol{E}$ and $\boldsymbol{H}$ in the radiation zone, eqs (6.59) show that $\boldsymbol{E}$ lies in a plane through the polar axis whereas $\boldsymbol{H}$ is azimuthal. The ratio

$$\frac{E}{H} = \left(\frac{\mu_0}{\epsilon_0}\right)^{1/2}$$

is just as for a plane electromagnetic wave *in vacuo* and since the fields are in phase all the characteristics of electromagnetic radiation *in vacuo* are present. That the hertzian dipole radiates electromagnetic radiation can be verified by calculating the flux of the Poynting vector $\boldsymbol{E}\times\boldsymbol{H}$ through a spherical surface centred on the dipole. Some care is needed when evaluating this quantity: the physical fields $\boldsymbol{E}$ and $\boldsymbol{H}$ are the *real* parts of fields such as eq. (6.22), and so vary as $\cos\omega[t]$. Thus

$$|S| \propto E_0H_0\cos^2\omega[t]$$

and the average Poynting vector over a complete cycle is

$$|S_{\mathrm{av}}| = E_0H_0\langle\cos^2\omega[t]\rangle = \tfrac{1}{2}E_0H_0.$$

The full result, which applies to the near and to the radiation zones for $\langle S\rangle_{\mathrm{av}}$ is

$$S_{\mathrm{av}} = \frac{\omega^4}{32\pi^2\epsilon_0 c^3}\frac{|\boldsymbol{r}\times\boldsymbol{p}_0|^2}{r^5}\boldsymbol{r} \tag{6.60}$$

and it involves only the radiation terms. The energy flow is seen to be zero along the axis of this dipole, and the Poynting vector falls off as $1/r^2$ because both the $\boldsymbol{E}$ and the $\boldsymbol{H}$ fields fall off as $1/r$ in the radiation zone. The $1/r^2$

dependence means that energy is conserved, for the total flux of the Poynting vector through a sphere centred on the dipole is

$$w = \frac{\omega^4 p_0^2}{12\pi\epsilon_0 c^3}$$

and so varies only with the square of p and the fourth power of ω, not with r.

Examples

6.1 Prove that if $\boldsymbol{B} = \nabla \times \boldsymbol{A}$, then $\boldsymbol{A} + \nabla\phi$ also satisfies this equation, where ϕ is *any* differentiable scalar field.

6.2 If magnetic monopoles do exist, two of Maxwell's equations will require some modification before they can describe such a particle. Thus

$$\nabla \cdot \boldsymbol{B} = \varrho^* \qquad \nabla \times \boldsymbol{E} = -\frac{\partial \boldsymbol{B}}{\partial t} - \boldsymbol{J}^*,$$

where $\boldsymbol{J}^*$ is a magnetic current density and ϱ^* a volume distribution of magnetic monopoles. Write down the two remaining Maxwell equations and note the 'symmetry'. Show that

$$\nabla \cdot \boldsymbol{J}^* = -\frac{\partial \varrho^*}{\partial t}$$

and for any closed path C, where $\partial B/\partial t = 0$

$$\oint_C \boldsymbol{E} \cdot \mathrm{d}\boldsymbol{l} = -I^*,$$

where I^* is a magnetic current.

6.3 Show that Maxwell's equations in free space are invariant under the transformation

$$\boldsymbol{E}' = \boldsymbol{E} \cos\theta + c\boldsymbol{B} \sin\theta$$

$$\boldsymbol{B}' = -(\boldsymbol{E}/c)\sin\theta + \boldsymbol{B}\cos\theta.$$

Show also that the energy density $\frac{1}{2}\epsilon_0 E^2 + \frac{1}{2}B^2/\mu_0$ and the Poynting vector $\boldsymbol{E} \times \boldsymbol{H}$ are invariant under this transformation. Why are *mixtures* of electric and magnetic fields also solutions to Maxwell's equations? (hint, see Chapter 7).

6.4 A polarized plane electromagnetic wave travels in a positive direction parallel to the z-axis. The electric field strength is

$$E_x = \sin\frac{\omega}{c}(z - ct).$$

Find an expression for the magnetic field $\boldsymbol{H}$.

6.5 Write down Maxwell's equations and show that for a homogeneous linear isotropic non-conducting medium the electric field satisfies

$$\nabla^2 \boldsymbol{E} - \epsilon\mu \frac{\partial^2 \boldsymbol{E}}{\partial t^2} = 0.$$

Find the corresponding equation for $\boldsymbol{H}$ and demonstrate that:

(1) The fields propagate with speed $(\mu\epsilon)^{-1/2}$ in the direction of propagation.
(2) $\boldsymbol{E}$ and $\boldsymbol{H}$ are each perpendicular to the direction of propagation and to each other.
(3) $\boldsymbol{E}$ and $\boldsymbol{H}$ are in time phase.

Chapter 7

Relativistic considerations

7.1 Galilean transformations

The *special* theory of relativity deals with observations made by two different observers, one of whom is moving at a constant velocity with respect to the other. The *general* theory of relativity deals with the case where the motion is not uniform, but we do not treat general relativity in this text. Figure 7.1 shows the two cartesian coordinate systems to which we will refer throughout the discussion of special relativity. The axis systems are labelled 1 and 2 and axis system 1 is assumed to be moving with relative velocity v along the x-axis. Corresponding x, y and z axes are parallel in the two systems.

It is usual to refer to such typical coordinate systems as *inertial frames*, and according to newtonian mechanics an event witnessed by an observer in frame 1 is related to the event as witnessed in frame 2 by the equations

$$\left.\begin{aligned} x_1 &= x_2 + vt \\ y_1 &= y_2 \\ z_1 &= z_2 \end{aligned}\right\} \tag{7.1}$$

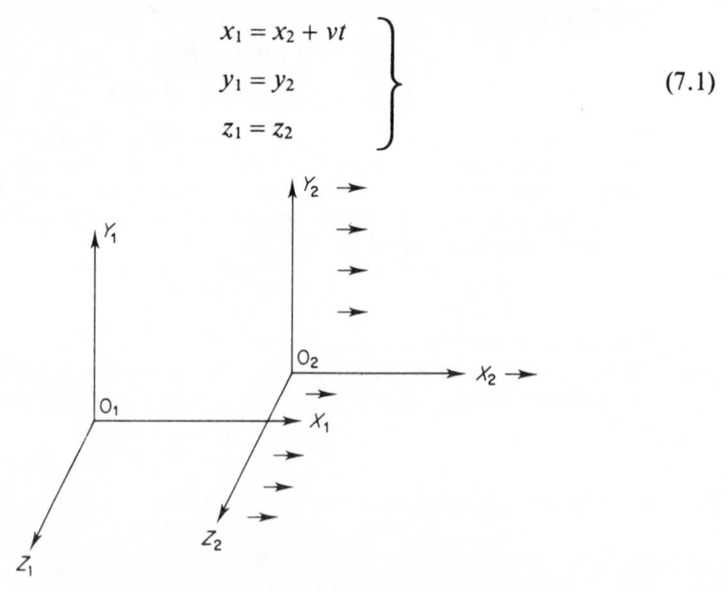

Fig. 7.1 Connection between the inertial frames

Equations (7.1) constitute a *galilean transformation*. The laws of classical mechanics look alike in all inertial frames and are said to be *invariant under galilean transformations*. Thus knowing a physical law in frame 1,

$$\boldsymbol{F}_1 = \frac{\mathrm{d}}{\mathrm{d}t}(m\boldsymbol{v}_1)$$

application of the galilean transformation (7.1) will give an identical law

$$\boldsymbol{F}_2 = \frac{\mathrm{d}}{\mathrm{d}t}(m\boldsymbol{v}_2)$$

where the subscripts refer to inertial frames 1 and 2. Implicit in the argument is the idea that time is an *absolute* quantity—the same to all observers in all inertial frames.

Two problems arise, however. First it is straightforward to demonstrate that the laws of electromagnetism are *not* invariant under galilean transformations, essentially because Maxwell's equations imply a fundamental limiting velocity, c. Secondly, both classical mechanics and the galilean transformation break down when velocities approach the speed of light. Thus, experiments with high energy fundamental particles in particle accelerators show that

(1) Particle velocities never exceed c no matter how high their energies. This means that the classical expression for the kinetic energy of a particle, $\frac{1}{2}mv^2$, must be incorrect.
(2) It is found that the resultant velocity of two velocities v_1 and v_2 is always *less* than $v_1 + v_2$, and in any case never exceeds c.
(3) Time intervals measured by observers in different inertial frames are different. This phenomenon is known as *time dilation*.

In order to overcome the difficulty associated with Maxwell's equations it was proposed that the laws of electromagnetism were only valid in one particular inertial frame called the *aether frame*. However, a famous experiment called the Michelson–Morley experiment which aimed to detect the motion of the earth through the aether gave a null result, and although the experiment has been repeated many times to higher degrees of experimental accuracy, a null result has always been produced. Thus if the validity of Maxwell's equations is accepted in one inertial frame it must be accepted in all inertial frames, and this suggests that the galilean transformation is in error.

7.2 The Lorentz transformation

Einstein's special theory of relativity is based on two postulates:

(1) The speed of light is the same in all inertial frames.
(2) It is not possible to detect the uniform motion of an inertial frame by measurements made entirely within that frame. This can be stated in the

alternative form that the laws of nature have the same appearance in all inertial frames.

The correct transformation relating laws in the two inertial frames of Fig. 7.1 is the *Lorentz transformation*

$$\left.\begin{aligned} x_1 &= \frac{x_2 + vt_2}{(1 - v^2/c^2)^{1/2}} \\ y_1 &= y_2;\ z_1 = z_2 \\ t_1 &= \frac{t_2 + (v/c)x_2}{(1 - v^2/c^2)^{1/2}} \end{aligned}\right\} \quad (7.2)$$

It is usual to write this more compactly by introducing

$$\beta = \frac{v}{c},\ \gamma = (1 - v^2/c^2)^{-1/2}$$

whence eqs (7.2) become

$$\left.\begin{aligned} x_1 &= \gamma(x_2 + vt_2) \\ y_1 &= y_2;\ z_1 = z_2 \\ t_1 &= \gamma(t_2 + \beta x_2) \end{aligned}\right\} \quad (7.3)$$

The Lorentz transformation, which has been verified experimentally, forms the basis of special relativity. The reverse transformation is obtained from eq. (7.3) by interchanging subscripts 1 and 2 and replacing v by $-v$, consistent with the physical picture that it is equally valid to regard Fig. 7.1 as showing inertial frame 2 at rest and inertial frame 1 moving in the $-x$ direction with speed v.

Two immediate consequences of the Lorentz transformation, eq. (7.3), are as follows. First, consider the case of observers in either frame equipped with identical clocks. If both observers note the time interval between periodic events in inertial frame 2 they will obtain different answers: if observer 2 records a time inverval Δt_2, observer 1 will record $\Delta t_1 = \gamma \Delta t_2 > \Delta t_2$, so the time interval will appear longer to the observer who is moving with respect to the event. This phenomenon, time dilation, has been observed in experiments on, for example, high-speed particles. Secondly, suppose that a ruler of length L is fixed along the x_2-axis such that its x_2 coordinates are 0 and L. An observer in inertial frame 1 will measure the length of this ruler by noting simultaneously (as far as he is concerned) the x_1 coordinates of each end of the ruler and subtracting. He will find $L_1 = \Delta x_1$ and since

$$\begin{aligned} \Delta x_2 &= \Delta(x_1 - vt_1) \\ &= \gamma \Delta x_1 \end{aligned}$$

then $L = \gamma L_1$ or $L_1 = L/\gamma$.

Thus, according to the moving observer the length in the direction of motion of the ruler is reduced by a factor γ. This is known as the *Lorentz–FitzGerald contraction* because it was predicted independently by Lorentz and by Fitz-Gerald, before Einstein's theory, in order to account for the null result of the Michelson–Morley experiment.

Starting from eqs (7.3) it is a straightforward but tedious matter to obtain formulae for the transformation of quantities between the inertial frames of Fig. 7.1. For example, the velocities V_1 and V_2 in frames 1 and 2 of Fig. 7.1 are related by

$$\left.\begin{aligned} V_{1,x} &= \frac{V_{2,x}+v}{1+\dfrac{V_{2,x}\,v}{c^2}} \\ V_{1,y} &= \frac{V_{2,y}}{\gamma\left[1+\dfrac{V_{2,x}v}{c^2}\right]} \\ V_{1,z} &= \frac{V_{2,z}}{\gamma\left[1+\dfrac{V_{2,x}v}{c^2}\right]} \end{aligned}\right\} \tag{7.4}$$

7.3 Relativistic mass

Unlike time and length measurements as in Section 7.2, all measurements of the amount of matter present by different observers yields the same result m_0, *the rest mass*. However, in order to make the conservation of momentum law $\sum m_i v_i = 0$ hold in all intertial frames it is necessary to redefine the quantities m_i to take account of the relative velocity of the inertial frames. A dependence of mass on velocity can be derived by considering the perfectly elastic collision of a pair of identical particles labelled A and B, which are travelling parallel to the x-axes of Fig. 7.1. Suppose that their velocities are measured in inertial frame 1 to be $+v$ and $-v$. According to eq. (7.4) their velocities in inertial frame 2 will be

$$V_{2,\mathrm{A}} = \frac{+V-v}{1-\dfrac{vV}{c^2}}; \quad V_{2,\mathrm{B}} = \frac{-V-v}{1+\dfrac{vV}{c^2}} \tag{7.5}$$

which clearly reduces to the newtonian values $\pm V - v$ for $vV \ll c^2$. If $m_{2,\mathrm{A}}$ and $m_{2,\mathrm{B}}$ are the particle masses measured by an observer in inertial frame 2 and if M_2 is the total mass in that inertial frame

$$m_{2,\mathrm{A}} + m_{2,\mathrm{B}} = M_2 \tag{7.6}$$

in order that mass be conserved, in accord with experiment. Again, if momentum is to be conserved as required by experiment

$$m_{2,\mathrm{A}}V_{2,\mathrm{A}} + m_{2,\mathrm{B}}V_{2,\mathrm{B}} = -M_2 v, \tag{7.7}$$

where total linear momentum before the collision has been equated to the momentum at the collision, when both particles are at rest according to the observer in frame 1 but travelling with velocity $-v$ according to observer 2. Combining eqs (7.6) and (7.7) gives

$$\frac{m_{2,\mathrm{A}}}{m_{2,\mathrm{B}}} = -\frac{v + V_{2,\mathrm{B}}}{v + V_{2,\mathrm{A}}} \tag{7.8}$$

from which it can be deduced that

$$m_{2,\mathrm{A}} = \frac{m_0}{(1 - V_\mathrm{A}^2/c^2)^{1/2}} \tag{7.9}$$

with a corresponding result for $m_{2,\mathrm{B}}$, where m_0 is the rest-mass. More generally the relativistic mass of an object with rest mass m_0, moving at speed v with respect to any observer is

$$m = \frac{m_0}{(1 - v^2/c^2)^{1/2}} = \gamma m_0 \tag{7.10}$$

and in the limit as $v \to 0$, $m \to m_0$. It must be stressed that eq. (7.10) implies only that the m_i appearing in $\sum m_i \boldsymbol{v}_i = \boldsymbol{0}$ need to be modified to take account of the Lorentz transformation; the amount of *matter* present does not vary!

7.4 Relativistic energy

From eq. (7.10)

$$\begin{aligned} mc^2 &= m_0 \gamma c^2 \\ &= m_0 c^2 + \frac{1}{2} m_0 v^2 + \frac{3}{8} m_0 \frac{v^4}{c^2} + \dots \end{aligned} \tag{7.11}$$

by the binomial expansion. The term $\frac{1}{2}m_0v^2$ is the newtonian kinetic energy and m_0c^2 is referred to as the *relativistic energy*. The quantity $mc^2 - m_0c^2$ is often called the *relativistic kinetic energy*, reducing to the newtonian expression $\frac{1}{2}m_0v^2$ for speeds $v \ll c$.

7.5 Invariance of electric charge

The idea of invariance has already been introduced: apart from physical laws being invariant under Lorentz transformations, quantities such as rest mass m_0 which have the same value in all inertial frames play an important role in special relativity, and such quantities are given the special name '*invariants*'. Thus masses m_i and lengths *are not* invariants but the speed of light *is* an invariant. All available experimental evidence suggests that *electric charge* is also an invariant. (The idea of invariance is rather different from the idea of conservation; not only is the amount of electric charge the same in all inertial frames (invariance), but in a given inertial frame electric charge cannot be created nor destroyed (conservation).)

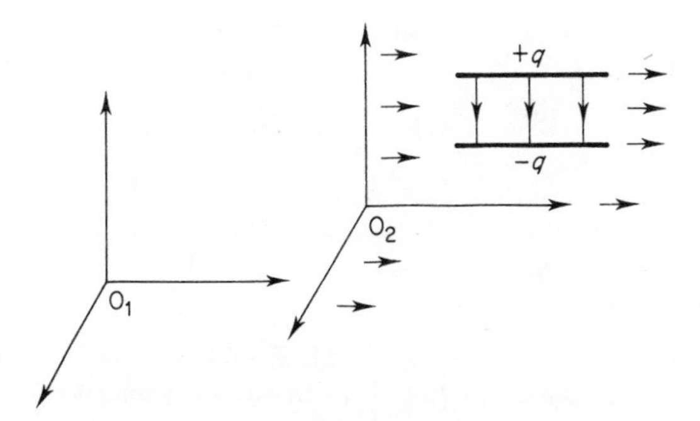

Fig. 7.2 Geometric construct used to illustrate the effect of motion on the length of a capacitor and field strength, as measured by observers in the two inertial frames

This fact, together with the Lorentz–FitzGerald contraction, suggests that electrostatic fields will appear differently to different observers. Thus Fig. 7.2 shows a parallel plate capacitor travelling at rest with respect to inertial frame 2 of Fig. 7.1.

For an observer in inertial frame 2

$$E_2 = \sigma_2/\epsilon_0$$

but for an observer in inertial frame 1, the length of the capacitor is reduced by a factor γ in the x direction and so, because of the invariance of electric charge, the charge density measured by the observer in frame 2 is greater by a factor γ and thus

$$E_1 = \gamma E_2. \tag{7.12}$$

7.6 Transformation of electric and magnetic fields

The Lorentz force for a charged particle moving with uniform

$$\boldsymbol{F}_1 = q(\boldsymbol{E}_1 + \boldsymbol{v} \times \boldsymbol{B}_1)$$

velocity $\boldsymbol{v}$ along the x-axis in Fig. 7.1 refers to measurements made by an observer in inertial frame 1. Assuming for the moment that $v \ll c$, an observer in inertial frame 2 travelling with the same velocity as the charge would see a *purely electrostatic* field

$$\boldsymbol{E}_2 = \boldsymbol{E}_1 + \boldsymbol{v} \times \boldsymbol{B}_1 \tag{7.13}$$

and so the magnetic induction $\boldsymbol{B}_1$ of inertial frame 1 has become an electrostatic field $\boldsymbol{v} \times \boldsymbol{B}_1$ in inertial frame 2. This suggests that the distinction between electric and magnetic fields is rather arbitrary. The general expressions relating $\boldsymbol{E}$ and $\boldsymbol{B}$ between the two inertial frames are

$$\left.\begin{aligned} \boldsymbol{E}_{2,\perp} &= \gamma(\boldsymbol{E}_{1,\perp} + \boldsymbol{v}\times\boldsymbol{B}_{1,\perp}) \\ \boldsymbol{E}_{2,\parallel} &= \boldsymbol{E}_{1,\parallel};\ \boldsymbol{B}_{2,\parallel} = \boldsymbol{B}_{1,\parallel} \\ \boldsymbol{B}_{2,\perp} &= \gamma\left(\boldsymbol{B}_{1,\perp} - \frac{\boldsymbol{v}\times\boldsymbol{E}_{1,\perp}}{c^2}\right) \end{aligned}\right\} \tag{7.14}$$

where the $\parallel$ and $\perp$ subscripts refer to components that are either parallel or perpendicular to the x-axis of Fig. 7.1. In the non-relativistic limit of $v \ll c$, eqs. (7.14) reduce to

$$\boldsymbol{E}_{2,\perp} = \boldsymbol{E}_{1,\perp} + \boldsymbol{v}\times\boldsymbol{B}_{1,\perp}$$

$$\boldsymbol{B}_2 = \boldsymbol{B}_1.$$

7.7 Four-vectors

Under a galilean transformation, the scalar distance $r = |\boldsymbol{r}_A - \boldsymbol{r}_B|$ between the points A, B retains the same value and so r is an invariant. This is not the case, however, with the Lorentz transformation, essentially because space and time are involved in the Lorentz transformation on an equal footing: an event in any particular inertial frame is characterized by the position of the event *and the time at which it occurs.* By analogy with three-dimensional geometry we define the coordinates of an event in space-time by the four-dimensional vector

$$\mathbf{R} = (x, y, z, ict) \tag{7.15}$$

and the scalar product is defined to be

$$\mathbf{R}\cdot\mathbf{R} = x^2 + y^2 + z^2 - c^2t^2.$$

It is readily verified from eqs (7.2) that

$$x_1^2 + y_1^2 + z_1^2 - c^2t_1^2 = x_2^2 + y_2^2 + z_2^2 - c^2t^2$$

and so $\mathbf{R}\cdot\mathbf{R}$ is an invariant. Note, however, that $\mathbf{R}\cdot\mathbf{R}$ can be negative on account of the $-c^2t^2$ term.

Any set of four quantities which transform in a manner similar to the components of $\mathbf{R}$ are said to form a *four-vector*. Four-vectors have the special properties that

(1) Their 'lengths' are invariants: thus if $\mathbf{W} = (w_1, w_2, w_3, w_4)$ is a four-vector then $\mathbf{W}\cdot\mathbf{W} = \sum w_i^2$ is an invariant.
(2) The scalar product $\mathbf{U}\cdot\mathbf{W} = \sum u_i w_i$ of two four-vectors is also an invariant.

In particle dynamics two other important four-vectors are the four-momentum

$$\mathbf{P} = (p_x, p_y, p_z, iE/c) \tag{7.16}$$

and the operator quad

$$\Box \equiv \left(\frac{\partial}{\partial x}, \frac{\partial}{\partial y}, \frac{\partial}{\partial z}, \frac{1}{ic}\frac{\partial}{\partial t}\right) \tag{7.17}$$

and hence both $p^2 - E^2/c^2$ and the d'Alembertian $\nabla^2 - 1/c^2\partial^2/\partial t^2$ are invariants.

7.8 Invariance of Maxwell's equations

As far as electromagnetism is concerned, we discussed in Section 7.6 the behaviour of the $\boldsymbol{E}$ and $\boldsymbol{B}$ fields under Lorentz transformations. The complicated form of the transformation laws mirrors the fact that $\boldsymbol{E}$ and $\boldsymbol{B}$ are not themselves components of four-vectors. It turns out to be more profitable to disucss the behaviour of Maxwell's equations under Lorentz transformations starting from the equivalent formulation of the equations in terms of the potentials $\boldsymbol{A}$ and V derived in Section 6.8:

$$-\nabla^2 V + \frac{1}{c^2}\frac{\partial^2 V}{\partial t^2} + \varrho/\epsilon_0 \tag{7.18}$$

$$-\nabla^2 \boldsymbol{A} + \frac{1}{c^2}\frac{\partial^2 \boldsymbol{A}}{\partial t^2} = \mu_0 \boldsymbol{J}. \tag{7.19}$$

In arriving at eqs (7.18) and (7.19) we have used the Lorentz conditions

$$\nabla \cdot \boldsymbol{A} = -\frac{1}{c^2}\frac{\partial V}{\partial t}. \tag{7.20}$$

The equation of continuity

$$\nabla \cdot \boldsymbol{J} + \frac{\partial \varrho}{\partial t} = 0 \tag{7.21}$$

was discussed in Section 6.2 because of its significance regarding the conservation of electric charge. Thus eq. (7.21) must surely be invariant under Lorentz transformations:

$$\nabla_1 \cdot \boldsymbol{J}_1 + \frac{\partial \varrho_1}{\partial t_1} = \nabla_2 \cdot \boldsymbol{J}_2 + \frac{\partial \varrho_2}{\partial t_2} = 0$$

or

$$\Box \cdot \mathbf{J} = 0,$$

where the four-current $\mathbf{J}$ is

$$\mathbf{J} = (J_x, J_y, J_z, i\varrho/c). \tag{7.22}$$

The invariant nature of eq. (7.21) suggests that $\mathbf{J}$ is a four-vector and this is indeed the case. Corresponding arguments show that the four-potential is also a four-vector

$$\mathbf{A} = (A_x, A_y, A_z, iV/c). \tag{7.23}$$

Thus the Lorentz conditions eq. (7.20) can be written in invariant form

$$\Box \cdot \mathbf{A} = 0$$

and eqs (7.18) and (7.19) can *all* be written

$$\Box^2 A_\mu = -\mu_0 J_\mu, \tag{7.24}$$

where μ takes integral values from 1 to 4. The operator $\Box^2$ is a Lorentz invariant and since both sides of eq. (7.24) are four-vectors, they transform the same way under the Lorentz transformation. This shows (albeit indirectly in this simple treatment) that Maxwell's equations satisfy the conditions of special relativity.

It is more usual to discuss the effects of special relativity in terms of Cartesian tensors. This is somewhat outside the scope of the present text.

Examples

7.1 Show that $r^2 - c^2t^2$ is a Lorentz invariant.

7.2 Show that, for the uniform motion of Fig. 7.1, velocities in the two inertial frames are related by

$$v_{1,x} = \frac{v_{2,x} + v}{D}; \; v_{1,y} = \frac{v_{2,y}}{\gamma D}; \; v_{1,z} = \frac{v_{2,z}}{\gamma D}$$

$$\text{where } D = \left[1 + \frac{v_{2,x}v}{c^2}\right]$$

and find the inverse relations for v_2.

7.3 Starting from the relative mass γm_0, show that the *relativistic kinetic energy* $mc^2 - m_0c^2$ reduces to $\frac{1}{2}m_0v^2$ for speeds very much smaller than c.

Further reading

A. F. Kip, *Fundamentals of Electricity and Magnetism*, McGraw-Hill Kogakusha International Student Edition, Tokyo, 1969.
I. S. Grant and W. R. Phillips, *Electromagnetism*, John Wiley and Sons Ltd, Chichester, 1978.
Paul Lorrain and Dale Corson, *Electromagnetic Fields and Waves*, W. H. Freeman & Co., San Francisco, 1970.
Electromagnetism, SM352, The Open University, Milton Keynes 1980

Part B
Electromagnetic properties

Chapter 8

Electrical properties

The basic principles of electric and magnetic behaviour have been treated in Part A. Now in Part B we begin to apply those principles to the electric and magnetic properties of molecules and systems of molecules. In this chapter we are concerned with charge distributions, electric fields and energies. Our approach is classical and phenomenological; the quantum-mechanical origin of these properties and their calculation will be treated in Chapter 15.

8.1 Multipole moments

Suppose we have a set of charges q_i at positions $\boldsymbol{r}_i$. From the results in Chapter 1 we can write down exactly the potential at a point $\boldsymbol{R}$ due to this distribution of charges: it is given by

$$4\pi\epsilon_0 V(\boldsymbol{R}) = \sum_i q_i / |\boldsymbol{R} - \boldsymbol{r}_i| , \tag{8.1}$$

that is, the sum of the potentials due to the individual charges, in accordance with the superposition principle. Although this expression is exact, it may sometimes be too detailed to be useful. If the charges are all much closer to one another than to the point $\boldsymbol{R}$, the potential will be determined mainly by some property of the arrangement of charge around an origin inside the distribution and the distance of $\boldsymbol{R}$ from this origin.

The inverses of the distances $|\boldsymbol{R} - \boldsymbol{r}_i|$ in eq. (8.1) are expanded about R to yield the Taylor series

$$\frac{1}{|\boldsymbol{R} - \boldsymbol{r}_i|} = \frac{1}{R} - \boldsymbol{r}_i \cdot \nabla \left(\frac{1}{R}\right) + \tfrac{1}{2}\boldsymbol{r}_i\boldsymbol{r}_i : \nabla\nabla \left(\frac{1}{R}\right) + \ldots \tag{8.2}$$

Here the second term on the right-hand side is a sum of terms like $x_i\partial(1/R)/\partial X$ as usual, and the third is a sum of terms like $x_i^2\partial^2(1/R)/\partial X^2$ and $x_iy_i\partial^2(1/R)\partial X\partial Y$ (see Appendix A.14). If eq. (8.2) is substituted in eq. (8.1), the potential is given as a sum of terms each of which is the product of two factors, one characteristic only of the charge distribution and the other characteristic only of the distance from the point $\boldsymbol{R}$:

$$4\pi\epsilon_0 V(\boldsymbol{R}) = \left(\sum_i q_i\right)\left(\frac{1}{R}\right) - \left(\sum_i q_i\boldsymbol{r}_i\right)\cdot\nabla\left(\frac{1}{R}\right)$$

$$+\tfrac{1}{2}\left(\sum_i q_i\boldsymbol{r}_i\boldsymbol{r}_i\right):\nabla\nabla\left(\frac{1}{R}\right)+\ldots \tag{8.3}$$

$$\equiv q\left(\frac{1}{R}\right) - \boldsymbol{p}\cdot\nabla\left(\frac{1}{R}\right) + \tfrac{1}{2}\mathbf{Q}:\nabla\nabla\left(\frac{1}{R}\right)+\ldots \tag{8.4}$$

This is a *multipole expansion* of the potential, in which the bracketed sums over the set of charges are the successive *multipole moments* of the distribution, namely the *total charge*

$$q = \sum_i q_i, \tag{8.5}$$

the *dipole moment* vector

$$\boldsymbol{p} = \sum_i q_i\boldsymbol{r}_i, \tag{8.6}$$

and the *quadrupole moment* tensor

$$\mathbf{Q} = \sum_i q_i\boldsymbol{r}_i\boldsymbol{r}_i. \tag{8.7}$$

Tensors are treated briefly in Appendix A.14.
Higher moments such as the octupole and hexadecapole moments are occasionally encountered, but need not be considered here. If the expansion is continued indefinitely, it always yields the exact potential. However, successive terms fall off more and more rapidly with R. Thus sufficiently far from the distribution the potential reduces to that of its lowest non-vanishing multipole moment.

The treatment of the total charge is straightforward, being governed by the equations of Chapter 1. Therefore although molecular systems carrying a net charge are of considerable importance, most obviously in ionic crystals, melts and solutions, they are not dealt with explicitly here.

The dipole moment of a pair of equal and opposite charges was introduced in Chapter 1. The calculation there showed explicitly how at distances large compared with the separation between the charges, the potentials of the charges tended to cancel, leaving a potential falling off as $1/R^2$ instead of $1/R$. This is the result implied by the first two terms on the right-hand side of eq. (8.3), where the general definition of the dipole moment (8.5) is used. One useful property of the dipole moment is that in a neutral system ($q = 0$) it is independent of the choice of origin. To show this, we suppose that a new origin is chosen at $\boldsymbol{r}_0$ relative to the old origin. The positions of the charges relative to the new origin are $\boldsymbol{r}_i' = \boldsymbol{r}_i - \boldsymbol{r}_0$, and the new dipole moment is therefore

$$\boldsymbol{p}' = \sum_i q_i \boldsymbol{r}_i' = \sum_i q_i \boldsymbol{r}_i - \sum_i q_i \boldsymbol{r}_0 \tag{8.8}$$

$$= \boldsymbol{p} - q\boldsymbol{r}_0 = \boldsymbol{p} \qquad \text{for } q = 0. \tag{8.9}$$

Thus tabulations of dipole moments of neutral molecules do not need to specify any choice of origin.

The quadrupole moment is a tensor described by a 3 × 3 matrix of components. The definition (8.7) shows that the matrix is symmetric, because for example

$$Q_{xy} = \sum_i q_i x_i y_i = \sum_i q_i y_i x_i = Q_{yx}. \tag{8.10}$$

Note that there are alternative ways of defining a quadrupole moment (see next section); the one in eq. (8.7) is simplest. The quadrupole moment is most important when the dipole moment is zero by symmetry, and in that case the quadrupole moment is also independent of origin (see the examples at the end of this chapter).

All the foregoing has been cast in terms of a discrete distribution of charges. For a continuous distribution of charge density $\varrho(\boldsymbol{r})$, the previous expressions are modified by replacing the charges q_i by elements of charge $\mathrm{d}q = \varrho(\boldsymbol{r})\,\mathrm{d}\tau$, where $\mathrm{d}\tau$ is an element of volume, and replacing the sum over i by an integral over $\mathrm{d}\tau$. Then eqs (8.5)–(8.7) become

$$q = \int \varrho(\boldsymbol{r})\,\mathrm{d}\tau \tag{8.11}$$

$$\boldsymbol{p} = \int \boldsymbol{r}\varrho(\boldsymbol{r})\,\mathrm{d}\tau \tag{8.12}$$

$$\mathbf{Q} = \int \boldsymbol{r}\boldsymbol{r}\varrho(\boldsymbol{r})\,\mathrm{d}\tau, \tag{8.13}$$

and eq. (8.4) remains unchanged in form.

8.2 Energy of charge distribution in a field

One way of studying charge distributions is to subject them to an electric field. This changes their energy and so can have an effect on spectroscopic properties.

Suppose that a charge distribution of the sort already considered is placed in an electrostatic potential which is not necessarily uniform in space. The energy of interaction is given exactly by

$$W = \sum_i q_i V(\boldsymbol{r}_i) \tag{8.14}$$

from the results in Chapter 1. As in the preceding section, we have an exact result which we manipulate in order to separate the properties of the charge distribution from those of the potential. We expand the potentials $V(\boldsymbol{r}_i)$ about the origin previously chosen, to obtain

$$V(r_i) = V + r_i \cdot \nabla V + \tfrac{1}{2} r_i r_i : \nabla \nabla V + \ldots \tag{8.15}$$

where V and its derivatives are all evaluated at the origin. Now we know that the electric field at the origin is

$$\boldsymbol{E} = -\nabla V \tag{8.16}$$

with the result that the electric field gradient is

$$\mathsf{E}' = \nabla \boldsymbol{E} = -\nabla \nabla V. \tag{8.17}$$

Substituting these results in $V(\boldsymbol{r}_i)$, we find

$$V(\boldsymbol{r}_i) = V - \boldsymbol{r}_i \cdot \boldsymbol{E} - \tfrac{1}{2} \boldsymbol{r}_i \boldsymbol{r}_i : \mathsf{E}' - \ldots, \tag{8.18}$$

which in eq. (8.14) yields the energy as

$$W = \left(\sum q_i\right) V - \left(\sum_i q_i \boldsymbol{r}_i\right) \cdot \boldsymbol{E} - \tfrac{1}{2}\left(\sum_i q_i \boldsymbol{r}_i \boldsymbol{r}_i\right) : \mathsf{E}' - \ldots, \tag{8.19}$$

We recognize the quantities in brackets as the multipole moments we met before, so that the energy can be written as

$$W = qV - \boldsymbol{p} \cdot \boldsymbol{E} - \tfrac{1}{2} \mathsf{Q} : \mathsf{E}' - \ldots. \tag{8.20}$$

Thus the multipole moments determine the interaction energy of the charge distribution with a varying potential, as well as the field due to the charge distribution. As eq. (8.20) shows, the number of multipole moments which contribute to the interaction energy depends on how rapidly the potential varies over the distribution. Fixing the spatial variation of the potential can thereby fix the multipole moment which determines the energy. In particular, in a uniform potential the energy is just qV as found in Chapter 1. In a uniform field the energy of a neutral charge distribution is

$$W = -\boldsymbol{p} \cdot \boldsymbol{E}, \tag{8.21}$$

a result we shall use several times. A corollary is that for constant field gradients and so on

$$\boldsymbol{p} = -\mathrm{d}W/\mathrm{d}\boldsymbol{E}. \tag{8.22}$$

Finally, if there is a field gradient, the quadrupole moment contributes to W; again, higher moments are not normally treated.

One special feature of the quadrupole moment term is that the components of the field gradient term are not all independent. The potential is taken at some point outside the distribution of external charges which produce it, and therefore satisfies Laplace's equation $\nabla^2 V = 0$, i.e.

$$E'_{xx} + E'_{yy} + E'_{zz} = 0. \tag{8.23}$$

The interaction energy is then unchanged if any constant is added to the diagonal elements Q_{xx}, Q_{yy} and Q_{zz} of the quadrupole moment tensor. This implies that, because the components of $\mathbf{E}'$ cannot all be varied independently, the variation of W with $\mathbf{E}'$ can only determine the diagonal elements of $\mathbf{Q}$ to within an arbitrary constant. It is therefore common practice to define a different quadrupole moment tensor which is traceless, with only five independent elements, all of which can be fixed by the variation of the energy with the field gradient:

$$\mathbf{\Theta} = \tfrac{1}{2} \sum_i q_i (3\mathbf{r}_i\mathbf{r}_i - r^2 \mathbf{1}). \tag{8.24}$$

Here $\mathbf{1}$ is the unit tensor, which in any axis system has its diagonal elements equal to 1 and its off-diagonal elements equal to 0; the factor $\frac{1}{2}$ is conventional.

8.3 Molecular multipole moments

The multipole moments can be determined experimentally through their interaction with electric fields. Dipole moments can be measured very accurately from the Stark effect on pure rotational (microwave) spectra of gases. In an electric field, the energy of a symmetric top rotor varies linearly with the dipole moment and the strength of the field, in a way which also depends on the three rotational quantum numbers. Application of the field then lifts the degeneracy of rotational transitions, giving splittings of the spectral lines. This is the first-order Stark effect, with splittings proportional to the field; the constant of proportionality yields the dipole moment. For linear molecules, there is no first-order Stark effect, but only a second-order effect which shifts spectral transitions by an amount proportional to the square of the field, again yielding the dipole moment. Dipole moments can be obtained this way to an accuracy of about 0.1 per cent. Alternatively, dipole moments can be determined from measurements of the low-frequency dielectric constant of gases or (less reliably) dilute solutions, by methods which are discussed in detail in the next chapter. Some dipole moments are given in Table 8.1. For comparison, transfer of one electron through a distance of 0.1 nm in a molecule produces a dipole moment of 16.0×10^{-30} C m. The old c.g.s. unit for dipole moment, the debye (D), corresponds to 3.336×10^{-30} C m.

Quadrupole moments are harder to determine experimentally because they require a non-uniform field. One method is to measure the deflection of a molecular beam in a field gradient. A few quadrupole strengths for axial molecules are given in Table 8.2. For such molecules the quadrupole tensor is diagonal in the molecular axes, with non-zero components $Q_{xx} = Q_{yy}$ and Q_{zz}.

Table 8.1 Selected permanent dipole moments

Molecule	Dipole moment/10^{-30} C m
CO	0.37
H_2O	6.2
SO_2	5.4
CH_3F	6.2
CH_3Cl	5.9
$CHCl_3$	3.5
C_6H_5Cl	5.7

Then the quadrupole strength defined as $Q = Q_{zz} - Q_{xx}$ completely determines the traceless quadrupole moment tensor discussed after eq. (8.23).

Molecular multipole moments determine how molecules tend to align, and hence affect the local structure in fluids and the packing in crystals. The energy of interaction of a pair of multipoles can be determined from the general expressions for the potential due to a charge distribution (8.4) and for the energy of a charge distribution in a field (8.20). For example, the energy of a dipole $\boldsymbol{p}_2$ in the potential due to a dipole $\boldsymbol{p}_1$ is given by

$$W = \boldsymbol{p}_1 \cdot [\nabla \nabla (1/R)] \cdot \boldsymbol{p}_2 / 4\pi\epsilon_0 \tag{8.25}$$

$$= [3(\boldsymbol{p}_1 \cdot \boldsymbol{R})(\boldsymbol{p}_2 \cdot \boldsymbol{R}) - R^2(\boldsymbol{p}_1 \cdot \boldsymbol{p}_2)]/4\pi\epsilon_0 R^5, \tag{8.26}$$

where $\boldsymbol{R}$ is the vector from dipole $\boldsymbol{p}_1$ to $\boldsymbol{p}_2$. In symmetrical situations, the more stable arrangements can often be seen by inspection. In Fig. 8.1 parts (a) and (b) show two idealized dipoles in the broadside position. Arrangement (b) is clearly the more stable since it has opposite charges nearer than like charges. In the axial position, parts (c) and (d), the energy is lower when the dipoles are parallel (d) than when they are antiparallel (c). Since like charges can be further apart in the axial position than in the broadside position for the same separation of centres, (d) is more stable than (b).

Table 8.2 Quadrupole strengths Q for selected axial molecules

Molecule	$Q/10^{-40}$ C m^2
H_2	2.2
N_2	−2.7
CO	−6.0
CO_2	−7.5
O_2	−1.3
F_2	2.9
C_6H_6 (benzene)	−29
C_6F_6 (hexafluorobenzene)	32

The interaction energy of two quadrupoles is more complicated, being given by

$$W = \mathbf{Q}_1 : [\,\nabla\,\nabla\,\nabla\,\nabla\,(1/R)] : \mathbf{Q}_2/4\pi\epsilon_0. \tag{8.27}$$

Parts (e)–(h) of Fig. 8.1 show some symmetrical arrangements of idealized quadrupoles. The broadside position (e) puts like charges together and so is less stable than the T position (g) where the double central charges are further apart. The same is clearly true if the sign of the quadrupole is changed, interchanging positive and negative charges. For two quadrupoles of opposite sign, however, the broadside position (f) is more stable than the T position (h). These deductions are nicely illustrated in benzene and hexafluorobenzene. In benzene the bonding tends to draw electrons in from the periphery of the ring, giving a quadrupole moment like that in (e). In hexafluorobenzene the strongly electronegative fluorine atoms tend to draw electrons out to the periphery, giving a quadrupole moment of opposite sign. Microwave spectroscopy of transient gas-phase dimers indicates that benzene and hexafluorobenzene alone

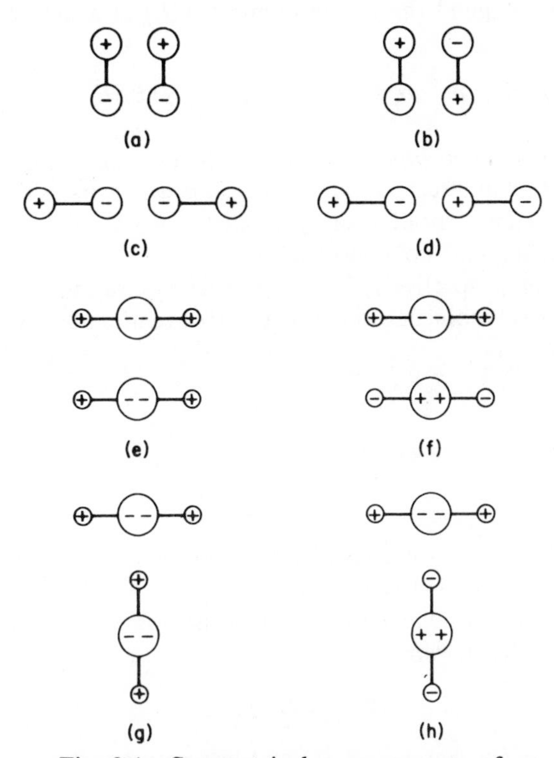

Fig. 8.1 Symmetrical arrangements of idealized dipoles (a–d) and quadrupoles (e–h). Arrangements (b), (d), (f) and (g) are the more stable

each give a dimer with a dipole moment, consistent with the T-shaped structure (g) or something similar but not with the broadside structure of parallel planes (e). However, mixed dimers of benzene-hexafluorobenzene show no measurable dipole moment, consistent with the broadside structure (f), where a dipole moment can develop only from charge redistribution in the dimer. There are similar indications of these preferred orientations in the liquid and crystal structures of the pure and mixed substances.

8.4 Induced dipoles

We have seen that the energy of a charge distribution changes in an electric field. If the charges can move, they will in general do so until the energy of the distribution in the field is minimized. This is the phenomenon of induced polarization already met in Chapter 2. In a molecule, the nuclei and the oppositely charged electrons will tend to move in opposite directions in the field, and the dipole will change.

Since the electric field perturbation is normally small and can be made arbitrarily so, we expand the dipole moment $\boldsymbol{p}(\boldsymbol{E})$ in a field $\boldsymbol{E}$ as a Taylor series:

$$\boldsymbol{p}(\boldsymbol{E}) = \boldsymbol{p}_0 + \boldsymbol{\alpha}\cdot\boldsymbol{E} + \frac{1}{2!}\boldsymbol{\beta}:\boldsymbol{EE} + \frac{1}{3!}\boldsymbol{\gamma}\vdots\boldsymbol{EEE} + \dots \tag{8.28}$$

Here $\boldsymbol{p}_0$ is the *permanent dipole moment*, $\boldsymbol{\alpha}$ is the *polarizability* tensor, and $\boldsymbol{\beta}$ and $\boldsymbol{\gamma}$ are the first and second *hyperpolarizability* tensors. Since $\boldsymbol{p}$ and $\boldsymbol{E}$ are both vectors, which are not necessarily parallel, $\boldsymbol{\alpha}$ is a second-rank tensor (see Appendix A.14). We shall not consider $\boldsymbol{\beta}$ and $\boldsymbol{\gamma}$ any further; they are third and fourth-rank tensors respectively. For the special case of an axially symmetric molecule, the dipole moment along the axis (conventionally labelled z) for a field along the axis is

$$p_z = p_0 + \alpha_{zz}E_z, \tag{8.29}$$

while perpendicular to the axis we have

$$p_x = \alpha_{xx}E_x, \tag{8.30}$$

since the symmetry requires $\boldsymbol{p}_0$ to lie parallel to the axis. The case of an electric field away from the axis is treated below.

The effect of induced polarization on the energy in a uniform electric field follows from eq. (8.22), which can be written as

$$\mathrm{d}W = -\mathrm{d}\boldsymbol{E}\cdot\boldsymbol{p}(\boldsymbol{E}). \tag{8.31}$$

Substitution for $\boldsymbol{p}(\boldsymbol{E})$ from eq. (8.28) and integration yields

$$W = W_0 - \boldsymbol{E}\cdot\boldsymbol{p}_0 - \tfrac{1}{2}\boldsymbol{E}\cdot\boldsymbol{\alpha}\cdot\boldsymbol{E} - \dots, \tag{8.32}$$

where W_0 is the energy in zero field. Hence the Cartesian components of the permanent dipole moment and polarizability can be obtained as

$$p_{0\alpha} = -(\mathrm{d}W/\mathrm{d}E_\alpha)_{E=0} \tag{8.33}$$

$$\alpha_{\alpha\beta} = -(\mathrm{d}^2W/\mathrm{d}E_\alpha \mathrm{d}E_\beta)_{E=0}. \tag{8.34}$$

The result (8.34) also shows that $\alpha_{\alpha\beta} = \alpha_{\beta\alpha}$ (by changing the order of differentiation) so that the 3×3 matrix of polarizability components is symmetric, leaving no more than six independent components.

From eq. (8.32) we can obtain the polarizability of an axial molecule with respect to a field at an angle θ to the axis. The contribution from the polarizability to W can be written in the molecular axes as

$$W' = -\tfrac{1}{2}(\alpha_{xx}E_x^2 + \alpha_{zz}E_z^2), \tag{8.35}$$

where the axial symmetry allows us to choose the x-axis so that the field has no y component. Now the energy is a scalar, independent of the choice of axis. We can therefore take the direction of $\boldsymbol{E}$ as one axis, in which case the energy is

$$W' = -\tfrac{1}{2}\alpha_\theta E^2, \tag{8.36}$$

where α_θ is the polarizability at the angle θ made by the field with the molecular axis z. Writing E_x and E_z as $E \sin\theta$ and $E \cos\theta$ respectively, we can equate these two expressions for W' to obtain

$$\alpha_\theta = \alpha_{xx}\sin^2\theta + \alpha_{zz}\cos^2\theta. \tag{8.37}$$

The appearance of the squares of trigonometric functions is characteristic of the angular transformation of second-rank tensors (see Appendix A.14).

8.5 Molecular polarizabilities

The experimental determination of the polarizability of a molecule is not

Table 8.3 Selected atomic and molecular mean polarizabilities α and polarizability anisotropies $\alpha_{zz} - \alpha_{xx}$ at optical frequencies

Species	$\alpha/10^{-40}$ F m^2	$(\alpha_{zz} - \alpha_{xx})/10^{-40}$ F m^2
He	0.23	0
Ne	0.44	0
Ar	1.83	0
H_2	0.90	0.35
N_2	1.97	0.77
CO	2.20	0.59
CO_2	2.93	2.34
CH_4	2.90	0
CH_3Cl	5.04	1.72
$CHCl_3$	9.5	−3.0
CCl_4	12.5	0
C_6H_6 (benzene)	11.6	−6.2

straightforward, especially if the molecule has little or no symmetry. As discussed in the next chapter, the average polarizability $(\alpha_{xx} + \alpha_{yy} + \alpha_{zz})/3$ can be determined from the refractive index or dielectric constant of a gas, and a measure of the anisotropy is given by Kerr effect measurements. A different measure governs the depolarization in light scattering. Polarizabilities can also be deduced from measurements on condensed phases, but here uncertainties in the theoretical treatments make the results somewhat unreliable. Representative polarizabilities are given in Table 8.3.

Molecular polarizabilities depend on the frequency of the electric field used in measuring them, being often higher for optical frequencies than for a static field. A simple model of the frequency dependence is presented in the next section, while the quantum-mechanical theory of polarizability is treated in Chapter 15.

8.6 Models of polarizability

An insight into the order of magnitude of polarizabilities is provided by a model which treats an atom as a point nucleus of charge $+Q$ at the centre of a uniform spherical distribution of negative charge of total charge $-Q$ and radius a. When an electric field E is applied, the nucleus is displaced a distance d from the centre of the charge cloud (see Fig. 8.2). At this point, the force QE on the nucleus due to the field is balanced by the force on the nucleus due to the negative charge. By Gauss's theorem, this is the same force as would be exerted if all the charge within a radius d were concentrated at the centre. This charge is $-Q$ multiplied by the ratio of the volumes of the spheres of radius d and a, i.e. $-Qd^3/a^3$. Acting at a distance d from the nucleus, this gives a force of magnitude $Q^2(d^3/a^3)/4\pi\epsilon_0 d^2 = Q^2 d/4\pi\epsilon_0 a^3$. Hence the

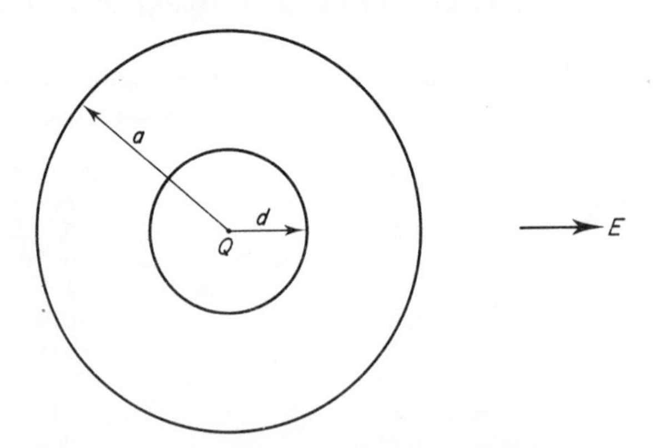

Fig. 8.2 A model of a polarizable atom. The electric field E displaces the nucleus of charge Q a distance d from the centre of the uniform spherical distribution of negative charge

displacement d satisfies

$$Q^2 d/4\pi\epsilon_0 a^3 = QE \tag{8.38}$$

But the induced dipole moment is Qd and the polarizability is therefore Qd/E, so that

$$\alpha = 4\pi\epsilon_0 a^3. \tag{8.39}$$

Thus apart from the dimensional factor $4\pi\epsilon_0$, α is determined by the cube of the size a of the atom, or roughly speaking by its volume.

The correlation between atomic or molecular size and polarizability is illustrated by the data in Table 8.3. The polarizability of the inert gas atoms increases in order of their size, and the mean polarizabilities of the molecules also tend to increase with molecular volume or number of electrons. The polarizability anisotropies can also be understood qualitatively in terms of the anisotropy of the molecular charge distribution. For the linear molecules, the polarizability is always larger along the axis than across it, because the charge distribution is always extended along the axis. For the axial molecules, the sign of the polarizability anisotropy depends on whether the charge distribution is predominantly extended along the axis or across it (egg-shaped or discus-shaped). Thus in CH_3Cl the readily polarizable chlorine atom on the axis makes α_{zz} larger than α_{xx}, while in $CHCl_3$ the readily polarizable chlorine atoms in the plane perpendicular to the axis make α_{xx} larger than α_{zz}.

The model above gives the restoring force on the displaced nucleus (or equivalently on the displaced electron distribution) as proportional to the displacement. It therefore corresponds to a simple harmonic oscillator of force constant $k = Q^2/4\pi\epsilon_0 a^3$. If the mass of the charge cloud is m and the nucleus is so much heavier that it can be treated as fixed, we can also write $k = m\omega_0^2$, where ω_0 is the radian frequency of the oscillator. Comparing these equations for k allows us to write the polarizability in terms of ω_0 as

$$\alpha = Q^2/m\omega_0^2. \tag{8.40}$$

These expressions give the polarizability in a static electric field. The polarizability of a harmonic oscillator of mass m carrying a charge Q in an alternating field of frequency ω is obtained from the equation of motion. The displacement $x(t)$ at time t satisfies the equation

$$m\frac{d^2x(t)}{dt^2} = QE(t) - kx(t), \tag{8.41}$$

where the right-hand side is the net force. At any time the induced dipole is $Qx(t)$, which is by definition equal to $\alpha(\omega)E(t)$, where $\alpha(\omega)$ is the polarizability at frequency ω. This allows us to substitute for $E(t)$ in terms of $x(t)$, and using $k = m\omega_0^2$ we find

$$m\frac{d^2x(t)}{dt^2} = \left[\frac{Q^2}{\alpha(\omega)} - m\omega_0^2\right]x(t). \tag{8.42}$$

We assume that $E(t) = E_0\, e^{i\omega t}$ and look for a solution of the form $x(t) = x_0\, e^{i\omega t}$ driven at the same frequency. Substitution in eq. (8.42) shows that this solution requires the polarizability to be given by

$$\alpha(\omega) = Q^2/m(\omega_0^2 - \omega^2). \tag{8.43}$$

This result includes the static polarizability of eq. (8.40). In this model, the polarizability increases as the frequency increases, as long as $\omega < \omega_0$, in agreement with the tendency referred to above.

From Chapter 6 we know that an oscillating dipole radiates electromagnetic waves. This damps the oscillation, causing the displacement to become out of phase with the driving field and requiring an input of energy to sustain the oscillation. These effects modify the polarizability. Dipole radiation depends on the current at time t, which in turn depends on the velocity $dx(t)/dt$. If on the right-hand side of eq. (8.42) we introduce a damping force, written for convenience as $-(m\gamma/\omega)\, dx(t)/dt$, where γ is a phenomenological damping constant, we find

$$\alpha(\omega) = \frac{Q^2/m}{\omega_0^2 - \omega^2 + i\gamma}. \tag{8.44}$$

Hence the polarizability becomes complex, its imaginary part indicating a component of induced dipole out of phase with the driving field. Quantum mechanics shows that molecular polarizabilities can be written as a sum of terms like (8.44), as treated in Chapter 15. The simpler classical expression derived in the present chapter is adequate for qualitative treatments of the frequency dependence of the polarizability and those dielectric and optical properties which it governs. Such properties form the subject matter of the next chapter.

Examples

8.1 Charges $-q$, $+3q$ and $-2q$ lie at positions $-a$, 0 and $+a$ along the x-axis.
  (a) Calculate the dipole and quadrupole moments about the origin.
  (b) Calculate the exact potential at positions $+3a$ and $+9a$ along the x-axis.
  (c) Calculate the potential given at the same points by the dipole and quadrupole moments calculated in (a).

  (Use $\partial r^{-1}/\partial x = -|x|/x^3$ and $\partial^2 r^{-1}/\partial x^2 = 2/x^3$.)

8.2 Show that the quadrupole moment of a charge distribution is independent of the choice of origin when the total charge and the dipole moment are both zero. (Work by analogy with eqs (8.8) and (8.9).)

8.3 A quadrupole consists of charges q at $\pm a$ along the z-axis and $-2q$ at the origin. Calculate the potential at a point in the xz plane a distance r from the origin by treating the quadrupole as equal and opposite point dipoles at $\pm \frac{1}{2}a$ along the z-axis, using the eq. (1.27). Expand to lowest order in a/r and show that the potential falls off more rapidly than that of either dipole alone.

8.4 Calculate the dipole moment induced in a molecule of polarizability 10×10^{-40} F m^2 by a potential difference of 1 kV applied across plates 1 cm apart.

Compare the result with a typical permanent dipole moment of 1 debye = 3.336×10^{-30} C m.

8.5 Evaluate the frequency dependence of the polarizability given by eq. (8.43). Plot $1/(1-\omega^2/\omega_0^2)$ for $\omega/\omega_0 = 0-2$ in steps of 0.2. What happens as $\omega \rightarrow \omega_0$?

8.6 Prove eq. (8.44).

8.7 Derive an expression for the real part of $\alpha(\omega)$ given by eq. (8.44). Plot the frequency dependence as in Example 8.5, assuming $\gamma = 0.2\omega_0^2$.

Further reading

R. Coelho, *Physics of Dielectrics*, Elsevier, Oxford, 1979.

C. J. F. Böttcher, *Theory of Electric Polarization*, Vol. I, Elsevier, Amsterdam, 1973.

C. J. F. Böttcher and P. Bordewijk, *Theory of Electric Polarization*, Vol. II, Elsevier, Amsterdam, 1978.

A. L. McLellan, *Tables of Experimental Dipole Moments*, Vol. I, Freeman, San Francisco, 1963; Vol. II, Rahara, El Cerrito, 1974.

D. E. Stogryn and A. P. Stogryn, *Molecular Physics* 11 (1966) 371. (Table of quadrupole moments.)

M. P. Bogaard and B. J. Orr, 'Electric dipole polarizabilities of atoms and molecules', in *MTP International Review of Science, Physical Chemistry, Series Two*, Vol. 2, Ch. 5, ed. A. D. Buckingham, Butterworth, London, 1975.

Chapter 9

Dielectric and optical properties

9.1 Introduction

The previous chapter dealt with properties of charge distributions from a rather general viewpoint. We now consider the electrical properties of assemblies of molecules. The results we shall obtain allow the bulk properties to be interpreted in terms of the molecular properties, and can also be used to determine the molecular properties. Dielectric and optical properties can be treated together because the refractive index n and the relative permittivity ϵ_r are related in non-magnetic insulators by

$$\epsilon_r = n^2, \tag{9.1}$$

as shown in Chapter 6. Experimentally it is found that ϵ_r and n are frequency dependent—for example, it is the frequency dependence of n in the optical region which gives rise to the familiar chromatic dispersion by which a prism produces a coloured spectrum from white light. The origin of this frequency dependence will be discussed shortly. The frequency dependence is often not explicitly indicated, so that it is important to remember that eq. (9.1) relates ϵ_r and n for the same frequency.

9.2 Gases

We begin by considering gases at low or moderate pressures. One mole of an ideal gas occupies 22.4 dm^3 under a pressure of 1 atm at 0 °C, whereas one mole of a molecular liquid occupies a volume typically of the order of 0.1 dm^3. The volume per molecule in the gas is thus some two hundred times the volume per molecule in the liquid, and the mean distance between molecules in the gas is consequently some six times that in the liquid. We are therefore able to neglect the interactions between the molecules in the gas—for example, the electric field of a dipole falls off as the cube of the distance and so would be two orders of magnitude smaller in the gas than in the liquid under the conditions discussed above.

Suppose we have N molecules of polarizability α in a volume V. For the time being, we assume that the molecules have no permanent dipoles, and that the polarizability is isotropic. When an electric field E is applied, each molecule acquires an induced dipole moment E. This gives a total moment $N\alpha E$ in a volume V, corresponding to a polarization

$$P = N\alpha E/V. \tag{9.2}$$

We therefore have a microscopic expression for P to compare with the macroscopic one we met in Chapter 2, namely

$$P = \epsilon_0 \chi_e E, \tag{9.3}$$

where χ_e is the electric susceptibility. We can then deduce that

$$\chi_e = N\alpha/\epsilon_0 V. \tag{9.4}$$

The relative permittivity is $1 + \chi_e$ (cf. Section 2.6), so that

$$\epsilon_r = 1 + N\alpha/\epsilon_0 V \tag{9.5}$$

and

$$n = (1 + N\alpha/\epsilon_0 V)^{1/2}. \tag{9.6}$$

In these expressions there occurs the combination $N\alpha/\epsilon_0 V$, which is inversely proportional to the volume per molecule. As we have seen, the volume per molecule is large in gases, so that this combination of factors is small, and ϵ_r and n differ little from unity. The right-hand side of eq. (9.6) can then be expanded by the binomial theorem to give

$$n \approx 1 + N\alpha/2\epsilon_0 V. \qquad \text{(dilute gas)} \tag{9.7}$$

For an ideal gas, N/V is p/kT, where p is the pressure, k the Boltzmann constant and T the absolute temperature, and the refractive index can be written as

$$n \approx 1 + \alpha p/2\epsilon_0 kT. \qquad \text{(dilute ideal gas)} \tag{9.8}$$

This expression provides a means of deducing α, by measuring n at various pressures and temperatures. If the molecules have different polarizabilities in different directions, this method yields the average polarizability.

From eq. (9.7) it can be seen that the frequency dependence of n follows from that of α. For a simple harmonic oscillator of frequency ω_0, the polarizability at a frequency ω varies like $1/(\omega_0^2 - \omega^2)$, as shown in the last chapter. The refractive index for a gas of such oscillators would then depend on frequency as shown in Fig. 9.1. As ω approaches ω_0 from below, α and hence n tend to $+\infty$; once ω exceeds ω_0, n increases from $-\infty$ to approach unity. In the presence of damping of the oscillator by some retarding force, the divergence is rounded off, as shown by the broken line. The quantum-mechanical theory of the polarizability in Chapter 15 shows that it consists of a series of terms with the same frequency dependence as the harmonic oscillator, with ω_0 replaced by the frequency corresponding to electronic transitions. The refractive index of a real gas thus shows a series of features

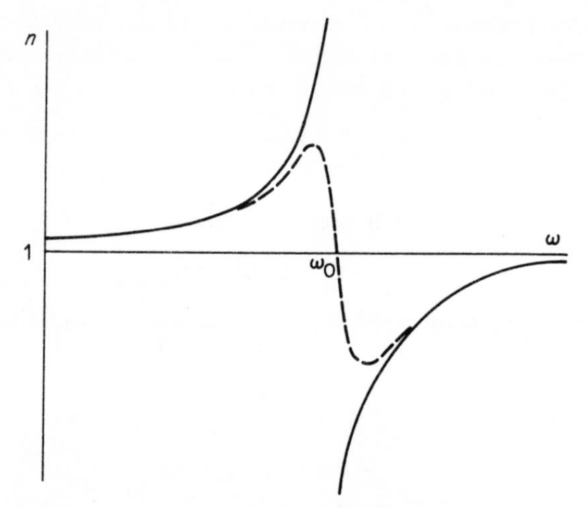

Fig. 9.1 Frequency dependence of the refractive index for an oscillator of frequency ω_0. The broken line shows the effect of damping

like that in Fig. 9.1 as the frequency sweeps through the molecular electronic transition frequencies, with damping attributable to the finite lifetime of the excited states (Chapter 13).

9.3 Condensed phases: the local field

In condensed phases (i.e. liquids and solids), the separation between molecules is of the order of molecular dimensions. It is then no longer possible to neglect the electrostatic interactions between the molecules. The result is that each molecule is polarized not by the ordinary macroscopic electric field E but by a *local field* F consisting of E plus the fields of all the other dipoles. Once the local field is known, the microscopic expression for the polarization, eq. (9.2), can be used, with the induced dipoles given by αF instead of αE. However, the calculation of F is complicated by the fact that the dipoles which contribute to F are themselves determined by F, so that a self-consistent treatment is necessary. In practice this is achieved by relating F to the polarization P and hence to the susceptibility χ_e, ultimately giving an equation for χ_e in terms of itself.

We may write the local field as

$$F = E + LP/\epsilon_0, \tag{9.9}$$

where L is the dimensionless *Lorentz factor,* which depends on the structure of the phase; ϵ_0 changes the dimensions from polarization to electric field. In general the local field will depend on the direction of E and P, so that L is strictly a tensor, and it can be shown that its three principal values always sum to unity. In cubic and isotropic phases, the three principal values must also be equal by symmetry, so that each equals 1/3. We then obtain the *Lorentz local field*

$$F = E + P/3\epsilon_0. \tag{9.10}$$

Substituting for P from eq. (9.3) permits F to be related to E by the equation

$$F = \left(1 + \frac{1}{3}\chi_e\right)E, \tag{9.11}$$

or in terms of the relative permittivity by

$$F = \frac{1}{3}(\epsilon_r + 2)E. \tag{9.12}$$

A method of deriving the Lorentz local field is treated in the examples at the end of this chapter.

The molecular expression for the polarization can now be obtained as $N\alpha F/V$, i.e.

$$P = (N\alpha/3V)(\epsilon_r + 2)E. \tag{9.13}$$

Alternatively, P is given by $\epsilon_0(\epsilon_r - 1)E$, so that equating the two expressions yields

$$\frac{\epsilon_r - 1}{\epsilon_r + 2} = \frac{N\alpha}{3\epsilon_0 V}, \tag{9.14}$$

which is the *Clausius–Mossotti* equation. At optical frequencies in non-magnetic materials we may replace ϵ_r by n^2 to obtain

$$\frac{n^2 - 1}{n^2 + 2} = \frac{N\alpha}{3\epsilon_0 V}, \tag{9.15}$$

which is the *Lorenz–Lorentz* equation. It can be seen that when $N\alpha/\epsilon_0 V$ is small, as in gases, ϵ_r and n^2 must be close to unity, so that $\epsilon_r + 2 \approx 3 \approx n^2 + 2$, and these equations reduce to those obtained above for gases.

The Clausius–Mossotti and Lorenz–Lorentz equations can be used to calculate ϵ_r and n from α or vice versa. It is convenient to rewrite their right-hand sides using

$$N/V = \varrho N_A/M, \tag{9.16}$$

where ϱ is the density, N_A the Avogadro constant and M the molar mass. Then we have

$$\frac{M}{\varrho}\frac{n^2 - 1}{n^2 + 2} = \frac{N_A\alpha}{3\epsilon_0} \equiv R_M, \tag{9.17}$$

which is the *molar refractivity*. This quantity depends only on the molecular polarizability and universal constants, and should therefore be independent of

Table 9.1 Molar refractivities for selected molecules and atoms

Molecule	n-C_6H_{14}	cyclo-C_6H_{12}	$(C_2H_5)_2O$	$(CH_3)_2CO$	$CHCl_3$
R/cm^3 mol^{-1}	29.8	27.7	22.5	16.2	21.4
Atom	C	H	=O	—O—	Cl
R/cm^3 mol^{-1}	2.4	1.1	2.2	1.6	6.0

temperature and pressure. Molar refractivities can be expressed quite well as a sum of atom or bond refractivities which are transferable between molecules, allowing refractive indices to be estimated for molecules of unknown polarizability. This possibility clearly depends on the existence of transferable atom or bond polarizabilities, which are of more fundamental interest than the refractivities. Some molar and atom refractivities are summarized in Table 9.1.

9.4 Orientation polarization

So far we have assumed that the molecules in the system have no permanent dipole moments. We now drop this restriction and allow the molecules to have a permanent dipole moment $\boldsymbol{p}_0$. A molecule with its dipole moment vector at an angle θ to an electric field has an energy

$$W = -\boldsymbol{p}_0 \cdot \boldsymbol{F} = -p_0 F \cos\theta. \tag{9.18}$$

The energy is lowest when $\theta = 0$, so that in a fluid the molecules will tend to orient themselves parallel to the field. This tendency will be opposed by random thermal agitation, but there will be some net orientation and hence a net *orientation polarization* $\boldsymbol{P}_0$, which can be calculated using methods of statistical mechanics.

By symmetry, $\boldsymbol{P}_0$ is parallel to $\boldsymbol{F}$ in a fluid sample, so that only the magnitude P_0 has to be calculated. This is given by

$$P_0 = N\langle p \rangle / V, \tag{9.19}$$

where $\langle p \rangle$ is the average dipole moment produced by the net orientation. All the dipoles have magnitude p_0, so that $\langle p \rangle$ is equal to $p_0 \langle \cos\theta \rangle$. The average value of $\cos\theta$ is obtained by its integral over θ weighted with (1) the geometrical probability of an angle lying between θ and $\theta + \mathrm{d}\theta$, and (2) the thermal probability of an energy W as given by eq. (9.18). The first probability is proportional to the area of a strip of radius $r \sin\theta$ and width $r\,\mathrm{d}\theta$ on the surface of a sphere of radius r (see Fig. 9.2), i.e. $2\pi r^2 \sin\theta\,\mathrm{d}\theta$. The thermal probability is proportional to the Boltzmann factor $\mathrm{e}^{-W/kT}$. Hence

$$P_0 = (Np_0/V)\langle \cos\theta \rangle \tag{9.20}$$

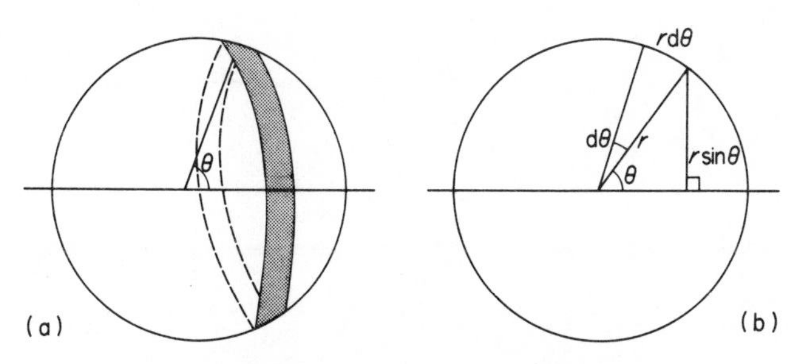

Fig. 9.2 (a) The geometrical probability of an angle θ with the axis is proportional to the area of the shaded strip on the surface of a sphere. (b) The strip has radius $r \sin \theta$ and width $r\, d\theta$

$$= (Np_0/V) \frac{\int_0^{\pi} \cos\theta\, e^{p_0 F \cos\theta/kT}\, 2\pi r^2 \sin\theta\, d\theta}{\int_0^{\pi} e^{p_0 F \cos\theta/kT}\, 2\pi r^2 \sin\theta\, d\theta}. \tag{9.21}$$

With the substitutions $u = p_0 F/kT$ and $x = \cos\theta$, this becomes

$$P_0 = (Np_0/V) \int_{-1}^{1} x e^{ux}\, dx \Big/ \int_{-1}^{1} e^{ux}\, dx. \tag{9.22}$$

The integrals can be evaluated by standard techniques to yield

$$P_0 = (Np_0/V)\mathcal{L}(u), \tag{9.23}$$

where $\mathcal{L}(u)$ is the *Langevin function*

$$\mathcal{L}(u) = \coth u - 1/u. \tag{9.24}$$

This function is sketched in Fig. 9.3. For large u, $\coth u \to 1$ and $1/u \to 0$, so that $\mathcal{L}(u)$ saturates at a value of unity; for small u, series expansion of coth u shows that $\mathcal{L}(u)$ varies as $u/3$.

In practice, values of the parameter u are no bigger than about 10^{-2} for realistic values of the other parameters, so that only the linear region of $\mathcal{L}(u)$ matters. The orientation polarization then becomes $Np_0 u/3V$, or

$$P_0 = (Np_0^2/3VkT)F, \tag{9.25}$$

so that a linear dependence on field is obtained. (Note that the Langevin function also applies to the orientation of magnetic dipoles.)

The orientation polarization given by eq. (9.25) must be added to the distortion polarization considered above, giving the total polarization

$$P = (N/V)[\alpha + p_0^2/3kT]F. \tag{9.26}$$

The derivation of the Clausius–Mossotti equation follows as before, except

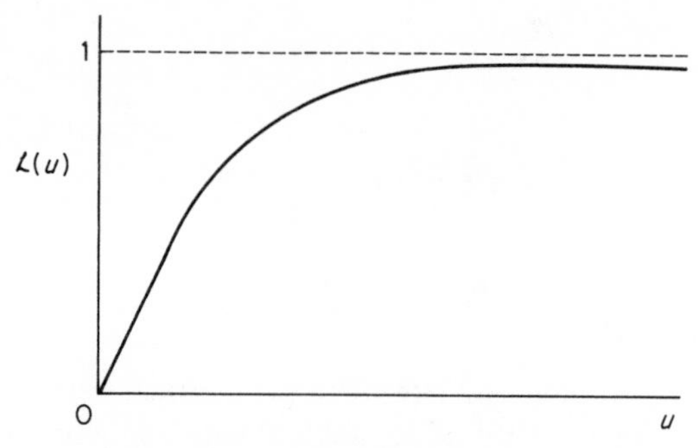

Fig. 9.3 The Langevin function, which behaves as $u/3$ for small u and tends asymptotically to unity as $u \to \infty$

that α is replaced by the combination in square brackets in eq. (9.26), so that with eq. (9.16) we obtain

$$\frac{M}{\varrho}\frac{\epsilon_r - 1}{\epsilon_r + 2} = \frac{N_A}{3\epsilon_0}[\alpha + p_0^2/3kT] \tag{9.27}$$

which is the *Debye equation*. This equation permits polarizabilities and dipole moments to be obtained from measurements of permittivity and density as a function of temperature (although the sign of the dipole moment is not determined because only p_0^2 matters). Reliable results are obtained for dilute gases and dilute solutions, but in more concentrated systems complications arise because the dipoles tend to associate through their electrostatic interaction. Furthermore, although the local field is taken into account in deriving the Debye equation, the Lorentz expression used ceases to be reliable because it does not take proper account of the orientation polarization; indeed, including this effect is a central feature of theories of polar liquids. Table 9.2 illustrates how the dipole moments deduced from the Debye equation depend on the conditions.

So far, the expressions we have derived have been valid at any frequency provided that parameters appropriate to that frequency are used. However, the Debye equation applies only at sufficiently low frequencies. The derivation

Table 9.2 Permanent dipole moments $p_0/10^{-30}$ C m determined under different conditions

Condition	Nitrobenzene	Phenylamine
Vapour	14.1	4.93
Carbon tetrachloride soution	13.1	4.87
Benzene solution	13.3	5.03
Carbon disulphide solution	12.2	4.73

assumes, in effect, that the orientation polarization responds instantaneously to the electric field. In fact, the molecules have a non-zero moment of inertia and can only reorient at a finite rate. Once the frequency of the electric field exceeds the characteristic rotational relaxation frequency, the orientation polarization is 'frozen out', and only the distortion component remains. Thus at optical frequencies the Debye equation is susperseded by the Clausius–Mossotti equation. If ϵ_r at low frequencies is denoted by ϵ_l and at high frequencies by ϵ_h, eq. (9.27) shows that $\epsilon_l > \epsilon_h$. The decrease in ϵ_r as the frequency is increased through the dipole relaxation frequency follows as

$$\epsilon_l - \epsilon_h = \frac{1}{9}(\epsilon_l + 2\epsilon_h)(\epsilon_h + 2)(Np_0^2/3\epsilon_0 VkT). \tag{9.28}$$

This difference is large in strongly polar liquids, where the large p_0 makes ϵ_l of the order of 10 or more; for example, in water at 25 °C ϵ_l is 79 and ϵ_h is 1.8.

A more detailed treatment of the frequency-dependent dielectric response of a molecular medium can be obtained as follows. The total polarization P is separated into two parts: a high-frequency part P_h due to electronic polarization and a frequency-dependent part P_f due to orientation polarization, which is much slower. It is assumed that p_h responds instantaneously to any applied electric field E, so that $P_h = \epsilon_0\chi_h E$; the frequency of the electric field is supposed to lie well below electronic excitation frequencies. On the other hand, it is assumed that P_f eventually reaches a value $\epsilon_0\chi_f E$ but relaxes towards this value at a finite rate proportional to the difference between the final and instantaneous values, i.e.

$$dP_f/dt = (\epsilon_0\chi_f E - P_f)/\tau. \tag{9.29}$$

Here τ is the *relaxation time* or inverse relaxation rate. For a constant field E switched on at time zero, P_f is found from eq. (9.29) to approach its final value exponentially as $e^{-t/\tau}$. For a field $E = E_0 e^{i\omega t}$, solution of eq. (9.29) yields

$$P_f = \epsilon_0\chi_f E/(1 + i\omega\tau). \tag{9.30}$$

Combining the results for P_h and P_f, we find that in a field of frequency ω the total polarization is given by

$$P = \epsilon_0[\chi_h + \chi_f/(1 + i\omega\tau)]E. \tag{9.31}$$

We can therefore define a complex relative permittivity

$$\epsilon(\omega) = 1 + \chi_h + \chi_f/(1 + i\omega\tau). \tag{9.32}$$

At high frequencies $\omega \to \infty$ the last term becomes negligible, leaving $\epsilon_h = 1 + \chi_h$ as we should expect. At low frequencies $\omega \to 0$ we obtain $\epsilon_l = 1 + \chi_h + \chi_f$, so that we can write χ_f as $\epsilon_l - \epsilon_h$. The complex relative permittivity is thus

$$\epsilon(\omega) = \epsilon_h + (\epsilon_l - \epsilon_h)/(1 + i\omega\tau); \tag{9.33}$$

in terms of relaxation, it is not ϵ_l itself which is significant but rather the difference $\epsilon_l - \epsilon_h$ due to orientation polarization.

It is customary to separate $\epsilon(\omega)$ into its real and imaginary parts. If we write $\epsilon(\omega) = \epsilon' - i\epsilon''$, then we find that the real part is

$$\epsilon' = \epsilon_h + (\epsilon_l - \epsilon_h)/(1 + \omega^2\tau^2) \tag{9.34}$$

and the imaginary part is

$$\epsilon'' = (\epsilon_l - \epsilon_h)\omega\tau/(1 + \omega^2\tau^2). \tag{9.35}$$

These are the *Debye equations* for a single relaxation process. The imaginary part is combined with a factor $-i$ in $\epsilon(\omega)$, rather than a factor $+i$ as usual in other contexts, because it is known that the orientation polarization must lag behind the applied field owing to the molecular inertia. As eq. (9.35) shows, the choice of $-i$ leads to a necessarily positive value of ϵ''. Since the electric field must do work to overcome the molecular inertia, ϵ'' is a measure of the loss of useful energy into heat, and may be referred to as the dielectric loss factor. The dissipation or absorption of energy is also measured by the loss angle δ such that $\tan \delta = \epsilon''/\epsilon'$; this represents the phase lag between the instantaneous polarization and the applied field, and measures the ratio between the energy lost and the energy stored per cycle.

The behaviour of ϵ' and ϵ'' as a function of frequency is illustrated in Fig. 9.4; note the logarithmic scale for $\omega\tau$. The real part ϵ' varies from ϵ_l at low frequency to ϵ_h at high frequency, changing most markedly in the region

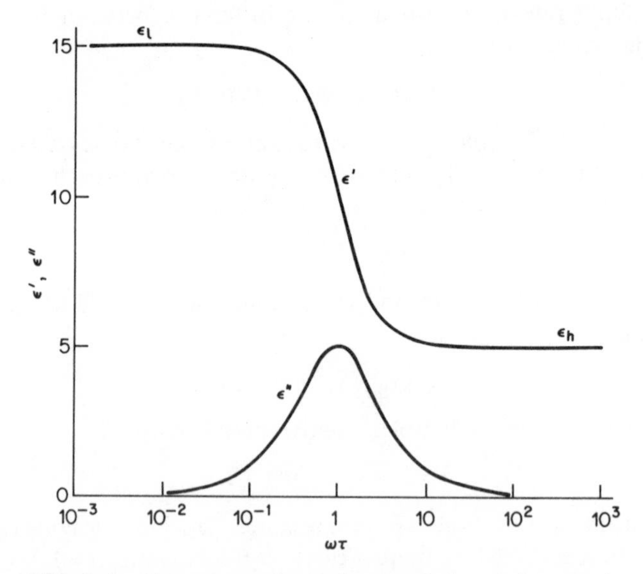

Fig. 9.4 Frequency dependence of the real part ϵ' and the imaginary part ϵ'' of the complex dielectric constant for a single Debye relaxation process, relaxation time τ. Here $\epsilon_l = 15$ and $\epsilon_h = 5$. Note the logarithmic scale for $\omega\tau$

$\omega\tau \approx 0.1-10$. The imaginary part ϵ'' is small at low and high frequency, being significant for $\omega\tau \approx 0.1-10$, with a maximum at $\omega\tau = 1$. How well experimental measurements of ϵ' and ϵ'' fit eqs (9.34) and (9.35) can be tested by combining them in the form

$$\epsilon''\omega = (\epsilon_l - \epsilon')/\tau \tag{9.36}$$

$$\epsilon''/\omega = (\epsilon' - \epsilon_h)\tau. \tag{9.37}$$

Plots of $\epsilon''\omega$ and ϵ''/ω against ϵ' should thus be linear, with slopes which both yield τ and intercepts which yield ϵ_l and ϵ_h. An alternative test is obtained by eliminating $\omega\tau$ from eqs (9.34) and (9.35) or more readily from eqs (9.36) and (9.37). This yields

$$(\epsilon'')^2 = (\epsilon_l - \epsilon')(\epsilon' - \epsilon_h), \tag{9.38}$$

which can be rearranged to the form

$$[\epsilon' - \tfrac{1}{2}(\epsilon_l + \epsilon_h)]^2 + (\epsilon'')^2 = [\tfrac{1}{2}(\epsilon_l - \epsilon_h)]^2. \tag{9.39}$$

This is the equation of a circle, showing that a plot of ϵ'' against ϵ' yields a semicircle (since ϵ'' takes only positive values) of radius $\frac{1}{2}(\epsilon_l - \epsilon_h)$ centred at $\frac{1}{2}(\epsilon_l + \epsilon_h)$. This is known as a *Cole–Cole plot*. Deviations from the Cole–Cole plot are usually ascribed to the presence of a distribution of relaxation times, and modified Cole–Cole plots have been devised to describe such deviations in terms of a single additional parameter characteristic of the distribution.

General theoretical arguments based on causality, in this case that the polarization must lag behind the field which produces it, show that ϵ' and ϵ'' are not completely independent. They must satisfy the *Kramers–Kronig* relations

$$\epsilon'(\omega) = \epsilon_h + \frac{2}{\pi}\int_0^\infty \frac{\epsilon''(x)x\,dx}{x^2 - \omega^2} \tag{9.40}$$

$$\epsilon''(\omega) = -\frac{2\omega}{\pi}\int_0^\infty \frac{[\epsilon'(x) - \epsilon_h]\,dx}{x^2 - \omega^2}, \tag{9.41}$$

each of which implies the other. As stated, these really refer to an upper limit of frequency such that ϵ' has attained the value ϵ_h and a lower limit such that it has attained the value ϵ_l. If instead the upper limit is taken above all electronic excitation frequencies, then ϵ_h is replaced by 1 and the dielectric loss processes include electronic absorption rather than just orientation polarization as assumed hitherto. The Kramers–Kronig relations then mean that it is possible to deduce the absorption spectrum implicit in $\epsilon''(\omega)$ by measuring the reflection spectrum which depends on $\epsilon'(\omega)$. This possibility is particularly useful for strongly absorbing crystals. However, in practice $\epsilon'(\omega)$ must be determined over a very wide range of frequency and special procedures must be adopted to avoid introducing spurious features into $\epsilon''(\omega)$ through truncating the infinite integration in eq. (9.41).

Dielectric relaxation is of practical importance because of its relation to energy loss and absorption. For example, the introduction of optical fibres for telecommunication has required the preparation of special glasses having exceptionally low losses at microwave frequencies. Not only must the glass itself have no relaxations near the relevant frequencies but it must also be free from all but the merest traces of impurities able to undergo dielectric relaxation in the same region. On the other hand, dielectric relaxation can provide valuable information on molecular motions in materials such as solutions and solid polymers. For example, amorphous polymers typically show what is referred to as an α–relaxation at high temperatures ascribed to movement of the polymer backbone, with a β-relaxation and possibly others at lower temperatures associated with movements of polar side groups. Study of the temperature dependence of the various loss peaks then allows activation energies to be deduced for the corresponding processes.

9.5 Kerr effect

The dielectric properties considered so far have mostly been linear ones, in which the response is directly proportional to the electric field (the exception was the general expression for the orientation polarization). We now turn to a non-linear property, the Kerr effect. Non-linear optics is a subject of considerable importance in modern optical technology, being essential to devices such as frequency doublers, used to convert laser light to light of double the frequency. It is also a subject of considerable complexity which lies largely beyond the scope of this book. However, the Kerr effect is fairly straightforward and is of importance both technically and because it provides a means of measuring the anisotropy of the molecular polarizability, as opposed to the average polarizability which enters the previous expressions.

The Kerr effect consists of the production of an *optical birefringence* in an isotropic fluid by a strong electric field applied transverse to the direction of propagation of the light; the refractive indices $n_{\parallel}$ and $n_{\perp}$ along and across the field then differ, and the magnitude of the difference is found to depend on the square of the electric field strength. An arrangement for measuring the Kerr effect is shown schematically in Fig. 9.5. Light from the source is polarized at 45° to the electric field applied across the Kerr cell. After passing through the cell, the light passes through a compensator and then through an analyser crossed with the original polarizer. The birefringence induced in the Kerr cell makes the light elliptically polarized. This ellipticity is nulled out by the compensator, the movement of which is proportional to the birefringence $n_{\parallel} - n_{\perp}$ and to the length of the Kerr cell. In practice, difficulties may arise from non-uniformity of the field and other factors. The sort of arrangement shown in Fig. 9.5 also finds use as a fast shutter. The birefringence is induced by the electric field as fast as the molecules can reorient, which may be on a nanosecond time-scale. This effect can be used in Q-switching a laser to produce intense pulses.

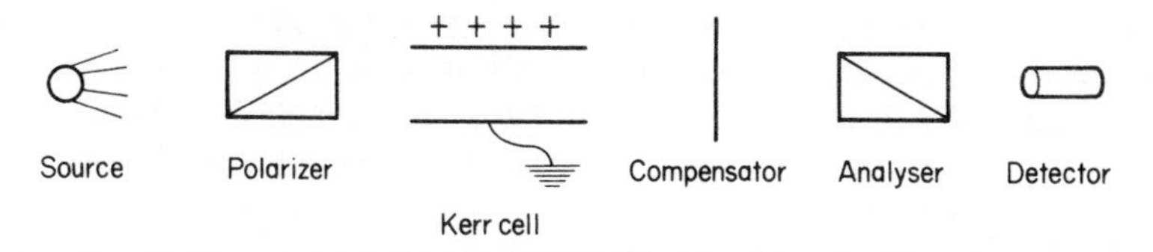

Fig. 9.5 Schematic arrangement for measuring the Kerr effect

The origin of the Kerr effect can be understood by considering for simplicity a dilute gas. Then we can use eq. (9.7) for the refractive index to write

$$n_{\parallel} - n_{\perp} = (N/2\epsilon_0 V)(\langle\alpha\rangle_{\parallel} - \langle\alpha\rangle_{\perp}), \tag{9.42}$$

where $\langle\alpha\rangle_{\parallel,\perp}$ denotes the thermal average polarizability in the specified direction. The effect of the electric field is to make these averages depend on direction. For a molecule with axial symmetry, the polarizability at an angle θ to the axis was given in the last chapter. This expression yields the difference in polarizability along and across a field at an angle θ to the axis as

$$\alpha_{\parallel} - \alpha_{\perp} = (\alpha_z - \alpha_x)\cos 2\theta. \tag{9.43}$$

The Kerr effect thus depends only on the polarizability anisotropy $\alpha_z - \alpha_x$ (where we recall that z and x are the molecular axes), and not on the absolute magnitude of the polarizability. It also depends on the thermal average $\langle\cos 2\theta\rangle$, which is calculated much as the orientation polarization was calculated but with the extra polarization energy term

$$W = \tfrac{1}{2}(\alpha_z\cos^2\theta + \alpha_x\sin^2\theta)E^2 \tag{9.44}$$

in the Boltzmann factor.

The difference in average polarizabilities is found to depend on E^2 to lowest order. A Kerr constant can then be defined as

$$K = (M/\varrho)(n_{\parallel} - n_{\perp})/E^2, \tag{9.45}$$

where the factor M/ϱ makes it a molar quantity (there are other ways of defining K which include various numerical factors). For the axially symmetric molecule in a very low pressure gas, calculation of $\langle\cos 2\theta\rangle$ yields

$$K = (N_A/30kT\epsilon_0)(\alpha_z - \alpha_x)[(\alpha_z - \alpha_x) + p_0^2/kT]. \tag{9.46}$$

Strictly speaking, the first factor $\alpha_z - \alpha_x$ refers to the optical frequency of the refractive index measurement and the second factor to the low frequency of the orienting field.

For non-polar molecules, a measurement of K gives the polarizability anisotropy directly, apart from sign; K is always positive. In this case the electric field orients the molecules through the polarizability. The direction of greatest polarizability gives the greatest energy lowering when parallel to the field, and so the molecules tend to orient this way.

For polar molecules, a measurement of K gives the polarizability anisotropy if the permanent dipole moment p_0 is known. Alternatively, both can be determined from measurements of K as a function of temperature. In this case the molecules tend to orient with their permanent dipoles parallel to the field. By symmetry, these moments must lie along the axis, so that the orientation tends to fix α_z parallel to the field. This allows K to be positive or negative, depending on whether α_z is larger or smaller than α_x (the permanent dipole term normally dominates the induced dipole term in eq. (9.46)). For molecules of lower than axial symmetry there is no single polarizability anisotropy and the expression for K is very complicated, but K is still positive or negative according as the largest polarizability is along or across the direction of the permanent dipole moment.

Thus, for example, in a molecule such as chlorobenzene where the direction of largest polarizability is parallel to the dipole moment, $K > 0$. In a molecule such as chloroform (trichloromethane) where the direction of largest polarizability is perpendicular to the dipole moment, i.e. in the plane of the chlorine atoms, $K < 0$. The Kerr effect can therefore be a source of structural information. A more complicated example is aniline (phenylamine), where the observation of a negative Kerr constant indicates that the $-NH_2$ group is unlikely to be coplanar with the benzene ring, since this would imply a large positive K as for chlorobenzene. These structural features and their relation

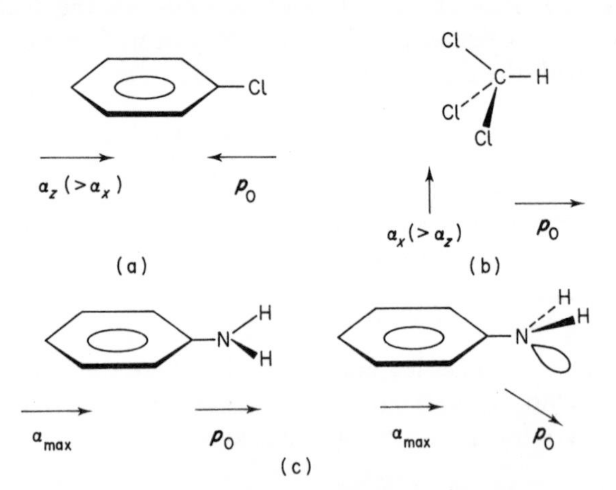

Fig. 9.6 The Kerr constant and molecular structure. (a) In chlorobenzene, $\alpha_{max} \parallel \mathbf{p}_0$ and $K > 0$. (b) In trichloromethane, $\alpha_{max} \perp \mathbf{p}_0$ and $K < 0$. (c) In phenylamine, a hypothetical planar structure has $\alpha_{max} \parallel \mathbf{p}_0$ and a positive K, but a non-planar struture allows the direction of p_0 to lie away from α_{max} to reduce the magnitude of K until it can become negative

Table 9.3 Molar Kerr constants for selected liquids and dilute gases

Liquid	CS_2	$CHCl_3$	C_6H_5Cl	$C_6H_5NH_2$	$C_6H_5NO_2$
$K/10^{-25}$ $m^5V^{-2}mol^{-1}$	25	−17	83	−10	2420
Gas	$(CH_3)_2O$	SO_2	$(CH_3)_2CO$	C_6H_6	C_2H_4
$K/10^{-25}$ $m^5V^{-2}mol^{-1}$	−1.3	−2.5	10	2.1	0.2

to the Kerr constant are illustrated in Fig. 9.6; some Kerr constants are given in Table 9.3.

9.6 Optical activity

Certain materials are found to rotate the plane of polarized light when it passes through them. In some crystals, such as quartz, this property disappears when the crystal is melted or dissolved, and so appears to be a property of the crystalline arrangement. In other materials composed of discrete molecules, the property persists in the crystal, the melt and the solution, and so appears to be a property of the molecular structure. The requisite structural feature is known to be *dissymmetry:* the crystal or molecular structure is not superimposable on its mirror image. The origin of this rule is discussed in Chapter 15; for the time being we are concerned with the nature of optical activity rather than its occurrence.

Optical activity can also be viewed as arising from the existence of different refractive indices for right-hand and left-hand circularly polarized light, or *circular birefringence.* Since plane polarized light can be viewed as a superposition of circularly polarized beams of opposite handedness (or chirality), this view is clearly equivalent to that first stated. It may seem more indirect but is in fact more fundamental, since the two spin states of a photon correspond to the two circularly polarized waves. The two waves also correspond to electric vectors describing helical paths of opposite handedness in space which one might imagine to be especially relevant to discriminating between structure of opposite handedness.

A circularly polarized electric field can be written as

$$\boldsymbol{E} = E_0(\boldsymbol{i}\cos\phi - \boldsymbol{j}\sin\phi), \tag{9.47}$$

where E_0 is the amplitude, $\boldsymbol{i}$ and $\boldsymbol{j}$ are unit vectors along the x and y axes, perpendicular to the direction of propagation of the wave, and ϕ is given by

$$\phi = \omega(t - nz/c). \tag{9.48}$$

Right- and left-hand polarizations are conveniently obtained by taking the frequency ω as positive or negative here, and n is understood to have a different value in each case. These relationships are illustrated in Fig. 9.7.

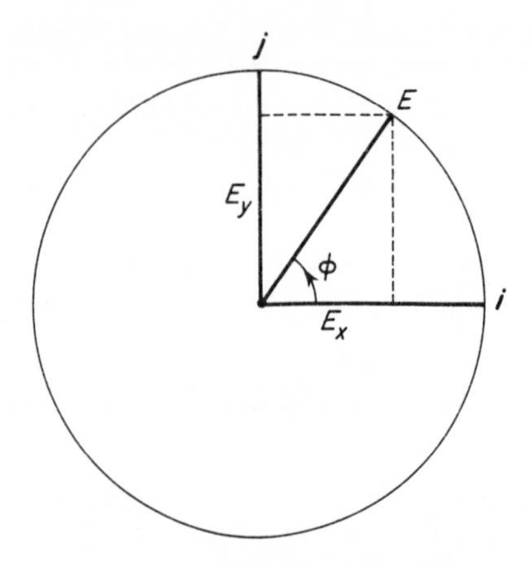

Fig. 9.7 Circular polarization. For propagation into the plane of the paper, right-handed polarization is given by ϕ changing in the sense shown

We assume that the optically active material is non-magnetic, so that $\mu_r = 1$. Then the refractive index is still given by the square root of the relative permittivity, which is still given by $1 + \chi_e$, but the electric susceptibility must now depend on the sign of the frequency, and hence so must the polarizability. It is also clear that in order to discriminate between molecules of opposite handedness it is necessary for the variation of the electric field over the molecule to be taken into account somehow instead of its value at one point. This implies that the spatial derivatives of $\boldsymbol{E}$ must be considered, which in an electromagnetic wave is conveniently achieved through curl $\boldsymbol{E}$. We therefore assume

$$\boldsymbol{p} = \alpha \boldsymbol{E} + \alpha'(\nabla \times \boldsymbol{E}). \tag{9.49}$$

In an electromagnetic wave we saw in Chapter 6 that E_z is zero while E_x and E_y are independent of x and y. This gives curl $\boldsymbol{E}$ as

$$\nabla \times \boldsymbol{E} = -\boldsymbol{i}\frac{\partial E_y}{\partial z} + \boldsymbol{j}\frac{\partial E_x}{\partial z}, \tag{9.50}$$

which for circular polarization yields

$$\nabla \times \boldsymbol{E} = -(\omega n/c)\boldsymbol{E}. \tag{9.51}$$

Thus we can write

$$\boldsymbol{p} = [\alpha - \alpha'(\omega n/c)]\boldsymbol{E}, \tag{9.52}$$

where α' can be seen as describing the response to $\omega \boldsymbol{E}$ or $\partial \boldsymbol{E}/\partial t$, and is thus proportional to the imaginary part of the total complex polarizability, while

α is the real part. Equation (9.52) has the dependence on ω required for optical activity to ensue.

The quantity in square brackets in eq. (9.52) is a new effective polarizability. This can be used in the dilute gas expression for ϵ_r (appropriate to the use of $\boldsymbol{E}$ rather than a local field in this section) instead of just α, to give

$$\epsilon_r = 1 + (N\alpha/\epsilon_0 V) - (N\alpha'/\epsilon_0 V)(\omega n/c). \tag{9.53}$$

The mean relative permittivity for right and left circular polarizations is

$$\bar{\epsilon}_r = 1 + N\alpha/\epsilon_0 V, \tag{9.54}$$

since the two polarizations correspond to equal and opposite frequencies. Hence

$$n = \epsilon_r^{1/2} = [\bar{\epsilon}_r - (N\alpha'/\epsilon_0 V)(\omega n/c)]^{1/2}, \tag{9.55}$$

where by writing $\bar{\epsilon}_r = \bar{n}^2$ and using the fact that $N\alpha'/\epsilon_0 V$ is small, we obtain

$$n \approx \bar{n}[1 - (N\alpha'/2\epsilon_0 V)(\omega n/c\bar{n}^2)]. \tag{9.56}$$

Finally we may set $n \approx \bar{n}$ on the right-hand side of this equation to find the two refractive indices in terms of the magnitude of ω:

$$n\pm = \bar{n} \mp N\alpha'\omega/2\epsilon_0 cV. \tag{9.57}$$

$$= \bar{n} \mp N\alpha'\omega\mu_0 c/V. \tag{9.58}$$

This result shows that the two waves have refractive indices differing by equal and opposite amounts from the mean refractive index. Since enantiomers (molecules of opposite handedness) have equal and opposite values of α', they produce equal and opposite rotations, and hence no rotation in an equimolar racemic mixture.

The rotation of plane polarized light is shown by writing it as the sum of right- and left-hand circularly polarized beams. Using eqs (9.47) and (9.48) with the refractive indices (9.58) one finds

$$\boldsymbol{E}_+ + \boldsymbol{E}_- = 2E_0 \cos\phi\ (\boldsymbol{i}\cos\delta\phi + \boldsymbol{j}\sin\delta\phi), \tag{9.59}$$

where the rotation $\delta\phi$ is given by

$$\delta\phi = (\omega z/c)\,\delta n = -(N/V)\omega^2\mu_0\alpha' z, \tag{9.60}$$

with δn the refractive index difference such that $n_\pm = \bar{n} \pm \delta n$. When δn is zero, pure x polarization ensues, but the difference between n_+ and n_- introduces a y component of polarization. As eq. (9.60) shows, the rotation so induced is proportional to the molecular concentration N/V and to the path length z. It is thus convenient to define a *molar rotation* per unit path length as

$$[\phi] = \delta\phi/z(n/V) \tag{9.61}$$

$$= -N_A\omega^2\mu_0\alpha', \tag{9.62}$$

Table 9.4 Specific rotation/deg kg^{-1} dm^{2} for selected substances in aqueous solution at 20 °C for the sodium D line

α-Glucose	β-Glucose	Sucrose	*d*-Tartaric acid
+112	+19	+66	+12
Quinine	*d*[Co en$_3$] Br$_3$.2H$_2$O		
−158	−117		

where n/V is the molar concentration (amount per unit volume). An older convention defines a *specific rotation* similarly in terms of the mass concentration m/V; it is therefore smaller by a factor M, the molar mass. Variations in M among related compounds (such as an optically active alcohol and its esters) may make the specific rotation vary much more than the molar rotation does, so that for purposes of comparison the molar rotation is preferable.

The molar or specific rotation of a compound is important in confirming its identity, in addition to other properties such as its melting point. For example, the optical activity is particularly useful for natural products such as carbohydrates. Because of the frequency dependence indicated by eq. (9.62), it is necessary to specify which light source is used in a measurement, often the sodium D line or the green mercury line. The temperature and the solvent may also have an effect. Illustrative values of specific rotation are given in Table 9.4.

As we have seen, optical activity can depend on the frequency of the light, both through the factor of ω^2 in eq. (9.62) and through any frequency dependence of α'. This dependence of optical rotation on frequency is called *optical rotatory dispersion* (ORD) and is particularly marked near molecular absorption frequencies (see Section 15.6). Optically active molecules also show the Cotton effect or *circular dichroism* (CD), that is, different absorption coefficients for circularly polarized light of opposite handedness, particularly near strong absorption lines. These effects are useful in establishing the relative stereochemical configurations of molecules.

Examples

9.1 Calculate the refractive index of an ideal gas of molecules of polarizability 1.00×10^{-39} F m^2 at a pressure of 1 atm and a temperature of 300 K.

9.2 The Lorentz local field can be calculated by considering a spherical region round a molecule large enough for the polarization outside to be treated as uniform and continuous. In cubic and isotropic materials the field due to the discrete array of molecules inside the spherical region averages to zero. The local field then differs from the macroscopic field by the field of the polarization surface charge density on the sphere. Express this charge density at an angle θ to the direction of the field in terms of the polarization P, and use Fig. 9.2 to show that the polarization field is $P/3\epsilon_0$.

9.3 Calculate the refractive index of a liquid of molecules of polarizability 1.00×10^{-39} F m^2 with molar volume 100 cm^3 mol^{-1}.
9.4 Verify that the Langevin function varies as $u/3$ to lowest order in u.
9.5 Calculate the value of the parameter $u = p_0E/kT$ for $p_0 = 5 \times 10^{-30}$ C m and $E = 10$ kV cm^{-1} at 300 K.
9.6 Evaluate the orientation polarization from eq. (9.22) to first order in u by using the series expansion for the exponential before integrating.
9.7 The temperature-dependent part of the molar refractivity of BrF_5 vapour is given by 1.428×10^{-2} m^3 mol^{-1} K/T at temperature T. Calculate the dipole moment of BrF_5.
9.8 The molar Kerr constant for hydrogen gas is 6.17×10^{-28} m^5 V^{-2} mol^{-1} at 300 K. Calculate the polarizability anisotropy $\alpha_z - \alpha_x$.
9.9 From the specific rotation for α glucose in Table 9.4 calculate the molar rotation and the quantity α'; take the relative molar mass (molecular weight) of glucose as 180 and the wavelength of the sodium D line as 589 nm.

Further reading

R. Coelho, *Physics of Dielectrics,* Elsevier, Oxford, 1979.
C. J. F. Böttcher, *Theory of Electric Polarization,* Vol. I, Elsevier, Amsterdam, 1973.
C. J. F. Böttcher and P. Bordewijk, *Theory of Electric Polarization,* Vol. II, Elsevier, Amsterdam, 1978.
A. R. Blythe, *Electrical Properties of Polymers,* Cambridge University Press, 1979.
J. H. Hannay, The Clausius–Mossotti equation: an alternative derivation, *Eur. J. Phys.,* **4** (1983) 141.

Chapter 10

Magnetic properties

10.1 Magnetic moment

In Chapter 3 we saw that isolated magnetic monopoles or 'charges' do not appear to exist. Instead, magnetism can be attributed to charged particles in motion, such as are observed in current electricity, or in general to charged particles possessing angular momentum. This generalization, which follows from relativistic quantum mechanics and is largely beyond the scope of this book, includes the magnetic effects of the 'internal' angular momentum known as *spin* as well as those of angular motion in space such as orbital angular momentum. As usual, we reserve discussion of quantum-mechanical aspects for Part C, where the magnetic moment is treated in Chapter 15. It is, however, convenient to take it as given that atoms and molecules have a net spin characterized by a quantity S which takes integer or half integer values and contributes to the magnetic moment in a manner to be described below.

The classical relationship between magnetic moment and angular momentum follows from the result in Section 3.3 showing that the magnetic induction on the axis of a coil of area A carrying a circumferential current I can be written at long distances as that of a magnetic dipole moment

$$\boldsymbol{m} = IA\boldsymbol{n}, \tag{10.1}$$

where $\boldsymbol{n}$ is the normal to the plane of the coil. Now for a particle of charge q moving with angular (radian) velocity ω, the current is $q\omega/2\pi$. If the particle has mass m and moves in a circle of radius r, the angular momentum $\boldsymbol{J}$ is parallel to $\boldsymbol{n}$ and has magnitude $mr^2\omega$. Substituting these results in eq. (10.1) using $A = \pi r^2$, we find that the magnetic moment and the angular momentum are related by

$$\boldsymbol{m} = (q/2m)\boldsymbol{J}. \tag{10.2}$$

The constant of proportionality for an electron is

$$\gamma = -e/2m_e, \tag{10.3}$$

which is the *magnetogyric ratio*, having the value -8.794×10^{10} Hz T^{-1}; here e is the proton charge, and the sign indicates that $\boldsymbol{m}$ and $\boldsymbol{J}$ are antiparallel.

As already noted, J will contain both orbital and spin contributions. In most molecules, however, the orbital contribution is zero (non-zero contributions require orbital degeneracy). The permanent magnetic moment due solely to the spin is then of magnitude

$$m_0 = g_e \mu_B [S(S+1)]^{1/2}, \tag{10.4}$$

where S is the spin quantum number. Here g_e is the *electron g-value*; its numerical magnitude is found by quantum electrodynamics to be 2.0023..., but it can often be approximated as just 2, and this will be done here. Also, μ_B is the *Bohr magneton*, equal to $|\gamma|\hbar$, and having the value 9.2737×10^{-24} J T^{-1}. Clearly μ_B has the dimensions of magnetic moment (note that J T^{-1} = A m^2), and it is convenient to express magnetic moments as so many Bohr magnetons.

Nuclei also have spin and hence magnetic moments. For a nuclear spin I, the magnetic moment is written as

$$m_0 = g_N \mu_N [I(I+1)]^{1/2}, \tag{10.5}$$

where g_N is the *nuclear g-value* and μ_N is the *nuclear magneton* $eh/2m_p$. Because μ_N contains the proton mass m_p instead of the much smaller electron mass in the Bohr magneton, μ_N is some 1800 times smaller than $\mu_B : \mu_N = 5.051 \times 10^{-27}$ J T^{-1}. Thus their greater mass means that nuclei have smaller magnetic moments than electrons, so that nuclear magnetism can normally be neglected compared with electronic magnetism. The exceptions

Table 10.1 Nuclear abundances, spins I and g factors g_N

Nucleus	% Abundance	I	g_N
^{1}H	99.98	$\frac{1}{2}$	5.59
^{2}H	0.02	1	0.86
^{12}C	98.9	0	—
^{13}C	1.1	$\frac{1}{2}$	1.40
^{14}N	99.6	1	0.40
^{15}N	0.4	$\frac{1}{2}$	−0.57
^{19}F	100	$\frac{1}{2}$	5.26
^{31}P	100	$\frac{1}{2}$	2.26
^{55}Mn	100	$2\frac{1}{2}$	1.38
^{59}Co	100	$3\frac{1}{2}$	1.33
^{63}Cu	69.1	$1\frac{1}{2}$	1.48
^{107}Ag	51.4	$\frac{1}{2}$	−0.23
^{127}I	100	$2\frac{1}{2}$	1.12
^{209}Bi	100	$4\frac{1}{2}$	0.90

mostly occur in magnetic resonance spectroscopy, which lies beyond the scope of this book. The sign and magnitude of the nuclear magnetic moment must be determined from experiment rather than from theory; some values are given in Table 10.1.

Measurements of atomic and molecular magnetic moments thus allow the spin S to be determined. Such measurements are of particular value in transition metal complexes, where they permit deductions to be made about the electronic configuration of the metal as modified by the ligands around it. Magnetic moments are obtained experimentally from magnetic susceptibilities, which are treated in the following section.

Magnetic multipole moments beyond the first are not normally encountered for molecular systems. This means that the magnetic induction produced by a molecule is usually written as that produced by the magnetic dipole moment alone, and the energy of a molecule of dipole moment $\boldsymbol{m}$ in a magnetic induction $\boldsymbol{B}$ is written as

$$W = W^0 - \boldsymbol{m} \cdot \boldsymbol{B}. \tag{10.6}$$

10.2 Magnetizability and susceptibility

Since the energy of a molecule changes in a magnetic induction, the charge distribution will distort in the induction so as to minimize the energy. For sufficiently small magnetic inductions, the response is linear and the magnetic moment is given by

$$\boldsymbol{m} = \boldsymbol{m}_0 + \boldsymbol{\varkappa} \cdot \boldsymbol{B}, \tag{10.7}$$

where $\boldsymbol{\varkappa}$ is the *magnetizability tensor*, analogous to the polarizability. The last term in eq. (10.7) is the induced magnetic moment. Using eqs (10.6) and (10.7) we can obtain the energy in a magnetic induction as

$$W = W^0 - \boldsymbol{B} \cdot \boldsymbol{m}_0 - \tfrac{1}{2} \boldsymbol{B} \cdot \boldsymbol{\varkappa} \cdot \boldsymbol{B} \tag{10.8}$$

(compare the derivation of eq. (8.32) for the energy of a polarizable molecule in an electric field). Unlike the polarizability, the magnetizability may be positive or negative. A sample with $\varkappa > 0$ is said to be paramagnetic, in accordance with the definition in Section 4.1, for when the term in $\boldsymbol{m}_0$ can be ignored the energy is lower in higher magnetic inductions, towards which the molecule is therefore attracted. Conversely, a sample with $\varkappa < 0$ is diamagnetic, being repelled from areas of higher magnetic induction, where its energy is higher.

The *magnetic susceptibility* tensor $\boldsymbol{\chi}_{\mathrm{m}}$ relates the magnetization or magnetic dipole density $\boldsymbol{M}$ to the magnetic field $\boldsymbol{H}$ as in Section 4.5, ignoring for the time being any permanent net magnetization:

$$\boldsymbol{M} = \boldsymbol{\chi}_{\mathrm{m}} \cdot \boldsymbol{H}. \tag{10.9}$$

The susceptibility is most useful at low fields, when it is independent of field. As discussed in Chapter 4, the magnetic induction is related to $\boldsymbol{H}$ and $\boldsymbol{M}$ via

$$\boldsymbol{B} = \mu_0(\boldsymbol{H} + \boldsymbol{M}), \tag{10.10}$$

so that

$$\boldsymbol{B} = \mu_0(\mathbf{1} + \chi_m) \cdot \boldsymbol{H} = \mu_0 \boldsymbol{\mu}_r \cdot \boldsymbol{H}, \tag{10.11}$$

where $\boldsymbol{\mu}_r$ is the relative permeability. In practice, $\boldsymbol{\mu}_r$ is very close to unity except for ferromagnets and other special materials, and correspondingly χ_m is much less than unity. Because the magnetization is so small, we can therefore take $\boldsymbol{B} \approx \mu_0 \boldsymbol{H}$ and ignore local field effects between molecules. (Local field effects *within* molecules are, however, important, giving rise for example to the chemical shift in nuclear magnetic resonance.) It is also customary to concentrate on the suceptibility rather than the relative permeability.

For N molecules with no permanent magnetic moment in a volume V, the magnetization is simply

$$\boldsymbol{M} = N\boldsymbol{\varkappa} \cdot \boldsymbol{B}/V = \mu_0 N\boldsymbol{\varkappa} \cdot \boldsymbol{H}/V \tag{10.12}$$

This applies even in a solid because there is no significant local field correction. Then the magnetic susceptibility is

$$\chi_m = \mu_0 N\boldsymbol{\varkappa}/V. \tag{10.13}$$

In particular this means that molecular magnetizabilities can be deduced from crystal, liquid, solution or gas susceptibilities without the uncertainties introduced in corresponding attempts to deduce polarizabilities from electric susceptibilities.

Magnetic susceptibilities can be measured by Gouy's magnetic balance, illustrated in Fig. 10.1. A cylindrical sample on one arm of a balance is suspended between the poles of an electromagnet. When the magnet is switched on, the sample is drawn into or repelled from the magnetic field. The change in the mass required to keep the sample balanced gives the sign and

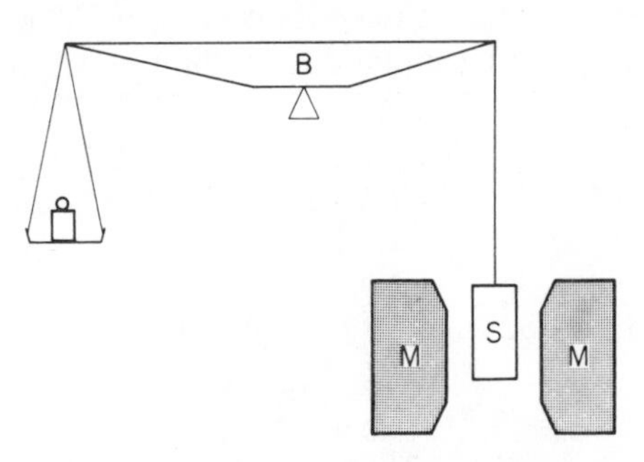

Fig. 10.1 Gouy magnetic balance (schematic). M = electromagnet, S = sample, B = balance beam

magnitude of the magnetic susceptibility. As shown in the examples at the end of this chapter, in an ideal case the mass difference varies as B^2. In practice, the balance is conveniently calibrated with samples of known susceptibility, for example, a solution of nickel(II) chloride. Representative susceptibilities are given in Table 10.2. The values given all refer to substances in their normal states at 300 K and 1 atmosphere. As eq. 10.13 shows, χ_m depends on the number of molecules per unit volume. Two other quantities related to the susceptibility may then be encountered in tables: the 'mass susceptibility' χ_m/ϱ and the 'molar susceptibility' $M\chi_m/\varrho$, where ϱ is the density and M the molar mass. The mass susceptibility is independent of the volume in which a given mass of substance is confined, and the molar susceptibility is independent of the mass as well. From eq. (10.13) it follows that $M\chi_m/\varrho$ is equal to $N_A\mu_0\varkappa$, so that the molar susceptibility is directly related to the molecular property $\varkappa$ and is hence best suited to comparisons between substances.

In fluid samples of molecules having permanent magnetic moments m_0, there is no net magnetization in the absence of a magnetic induction or field. However, in a magnetic induction, an orientation magnetization is produced. For the magnetic inductions normally attainable this magnetization is linear in B and has the form obtained in the last chapter for the orientation polarization at low field, namely

$$\boldsymbol{M}_0 = Nm_0^2\boldsymbol{B}/3VkT. \tag{10.14}$$

Substituting $\mu_0 H$ for B and using eq. (10.4) to express m_0 in terms of the spin S, we can obtain the scalar *spin paramagnetic susceptibility*

$$\chi_m^{\text{spin}} = 4N\mu_0\mu_B^2 S(S+1)/3VkT. \tag{10.15}$$

It is paramagnetic, being necessarily positive. This is the *Curie law*, $\chi_m \propto 1/T$, which holds provided T is not too low, when quantum effects become important. Empirically it is often found that a temperature dependence like

Table 10.2 Selected magnetic susceptibilities χ_m at 300 K and 1 atm, and corresponding molar susceptibilities $M\chi_m/\varrho$

Substance	χ_m	$(M\chi_m/\varrho)/\text{m}^3\ \text{mol}^{-1}$
He	-1.1×10^{-9}	-2.4×10^{-11}
H_2	-2.2×10^{-9}	-5.0×10^{-11}
N_2	-6.8×10^{-9}	-15.1×10^{-11}
O_2	$+1.9\times10^{-6}$	$+4.3\times10^{-8}$
NO	$+0.8\times10^{-6}$	$+1.8\times10^{-8}$
CO_2	-11.7×10^{-9}	-26.0×10^{-11}
H_2O	-9.0×10^{-6}	-16.2×10^{-11}
$NiCl_2$	$+27.9\times10^{-4}$	$+10.2\times10^{-8}$
$CuSO_4.5H_2O$	$+1.8\times10^{-4}$	$+1.9\times10^{-8}$
$MnSO_4.4H_2O$	$+17.1\times10^{-4}$	$+18.1\times10^{-8}$

$1/(T-A)$ fits the results; this is the Curie–Weiss law, with A the Curie temperature.

All molecules produce an orbital susceptibility χ_m^{orb} given by eq. (10.13). However, this is normally swamped by the spin susceptibility unless χ_m^{spin} is zero. The calculation of the magnetizability and hence of χ_m^{orb} is treated in Chapter 15, where it is shown that there are both paramagnetic and diamagnetic contributions, although the diamagnetic part usually dominates. Nevertheless, χ_m^{orb} is very small, typically of the order of 10^{-6}, as Table 10.2 shows.

10.3 Cotton–Mouton effect

In general, molecules are anisotropically magnetizable, so that even when they have no permanent magnetic moment their energy in a magnetic induction depends on their orientation. Such molecules therefore tend to align with their axis of algebraically largest magnetizability along the direction of the magnetic induction if they are paramagnetic or across it if they are diamagnetic. Alignment of this sort will also entail an alignment of the molecular polarizabilities and hence a magnetically induced optical birefringence. This is the essence of the *Cotton–Mouton effect*. It is analogous to the Kerr effect of Section 9.5, except that there the anisotropic polarizability causes both the alignment and the consequent birefringence.

The theory of the Cotton–Mouton effect is very similar to that of the Kerr effect. The lowest-order non-vanishing contribution varies as the square of the magnetic induction, leading to the definition of the molar Cotton–Mouton constant as

$$C = (M/\varrho)(n_{\parallel} - n_{\perp})/B^2. \tag{10.16}$$

Assuming again that the molecules are axial and have no permanent magnetic moment, one finds that C is given in terms of molecular properties by

$$C = (N_A/30kT\epsilon_0)(\alpha_z - \alpha_x)(\varkappa_z - \varkappa_x), \tag{10.17}$$

where α_z and $\varkappa_z$ are the polarizability and magnetizability along the molecular axis and α_x and $\varkappa_x$ are the corresponding quantities across the axis.

Measurement of the Cotton–Mouton effect requires a powerful magnet producing inductions of the order of a few T, and so is not frequently undertaken. As eq. (10.17) indicates, given the anisotropy of the polarizability or of the magnetizability, knowledge of C serves to determine the anisotropy of the other. This can be particularly useful to determine the complete polarizability tensor of a molecule of lower than axial symmetry, for which the mean polarizability is obtained from refractive measurements and another measure of the anisotropy from Kerr effect measurements; the anisotropic magnetizability is obtained from the crystal magnetic susceptibility as already noted.

10.4 Faraday effect

The Kerr and Cotton–Mouton effects both give rise to a birefringence quadratic in a field applied across the direction of propagation of light through the medium. In contrast, the *Faraday effect* is the production of circular birefringence by a magnetic induction along the direction of propagation. The effect is conveniently observed by passing plane-polarized light along the axis of a solenoid, as shown in Fig. 10.2. It is found that the rotation θ is proportional to the path length l in the magnetic induction and to the magnitude of the induction:

$$\theta = VlB, \tag{10.18}$$

where V is the *Verdet constant.*

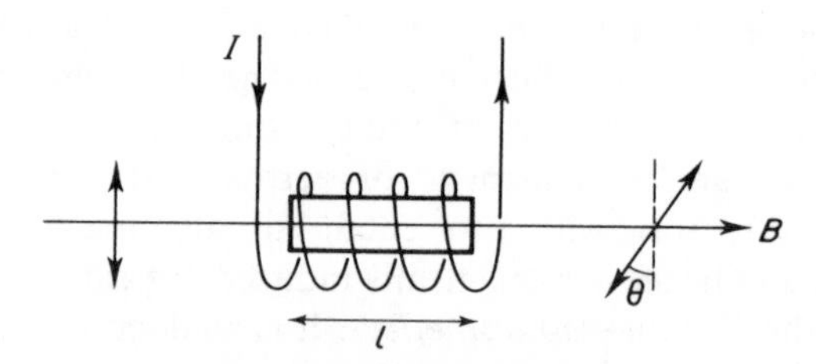

Fig. 10.2 The Faraday effect. Plane polarized light passing along the axis of a solenoid carrying a current I producing a magnetic induction B emerges with its plane of polarization rotated by an angle θ proportional to the sample length l

The origin of the Faraday effect can be understood with reference to the discussion of optical activity in Section 9.6. Rotation of the plane of polarized light arises from the molecular parameter α' related to the imaginary part of the polarizability. In optically inactive molecules α' is zero, but application of a magnetic induction produces a chirality in the molecules, yielding a non-zero α' and hence optical activity. The Verdet constant is thus related to the ease with which the molecule is distorted by the magnetic induction.

The Faraday effect may be referred to as *magnetic optical rotation* (MOR) and its dependence on frequency as *magnetic optical rotary dispersion* (MORD). There is also a linear magnetically induced analogue of circular dichroism, *magnetic circular dichroism* (MCD). These effects allow the techniques of ORD and CD to be extended to molecules which are not naturally optically active.

The Faraday effect is also studied in semiconductor crystals, where it yields information about the electronic energy band structure.

Examples

10.1 Calculate the magnetic moment of the hydrogen atom (^{1}H, $S = \frac{1}{2}$) in the ground electronic state, where orbital contributions are zero.

10.2 The component of the magnetic moment of the hydrogen atom along a magnetic induction is $\pm\mu_B$. Calculate the energy difference between these orientations in an induction of 0.1 T, and the frequency to which this energy corresponds. At what temperature T does the energy difference equal kT?

10.3 Show that the energy of magnetization is $-\frac{1}{2}V\chi_m B^2/\mu_0$, starting from eq. (10.8) for a single molecule.

10.4 In Gouy's magnetic balance (Fig. 10.1), a cylinder of cross-sectional area A of material of susceptibility χ_m is suspended between the poles of a magnet producing an induction B. The rest of the volume between the poles is filled with a fluid of susceptibility χ_0. Write down the energy of magnetization of the cylinder and fluid when the cylinder occupies a volume V out of the total volume V_0 between the poles. Work out the force on the cylinder by considering how this energy changes as the cylinder is withdrawn from between the poles along its axis. Hence show that the change in the mass required to balance the sample when the magnet is switched on is

$$\Delta m = \frac{1}{2}(\chi_m - \chi_0)AB^2/\mu_0 g,$$

where g is the acceleration due to gravity.

10.5 The susceptibility of water is -0.9×10^{-5} at 300 K. Deduce the mean magnetizability of the water molecule.

10.6 The susceptibility of $CuSO_4.5H_2O$ crystals is 1.76×10^{-4} at 300 K. Deduce the magnetic moment of the Cu^{2+} ion and the probable number of unpaired electrons (there is a small orbital contribution to χ_m). $CuSO_4.5H_2O$ has a molar mass of 250 g mol^{-1} and a density of 2.284 g cm^{-3}.

Further Reading

P. W. Atkins, *Molecular Quantum Mechanics*, 2nd edn, Clarendon Press, Oxford, 1983.

P. J. Wheatley, *The Determination of Molecular Structure*, 2nd edn, Clarendon Press, Oxford, 1968.

J. H. Van Vleck, *The Theory of Electric and Magnetic Susceptibilities*, Oxford University Press, 1965.

Part C
Quantum mechanics and electromagnetism

Chapter 11

Basic quantum mechanics

11.1 Introduction

In Parts A and B we described electromagnetism and electromagnetic properties without recourse to quantum mechanics. Indeed, Maxwell's equations represent a high point of classical physics, incorporating as they do the essential features of electricity, magnetism and optics in a relativistically invariant manner. However, the absorption and emission of radiation cannot be explained without introducing quantization of energy and photons. For a more fundamental understanding of these phenomena and of such quantities as dipole moments and polarizabilities, a proper quantum-mechanical treatment is needed. Therefore Part C combines the ideas of quantum mechanics with those of electromagnetism in order to interpret electromagnetic behaviour in terms of the quantum-mechanical states of molecular systems and transitions between them.

The present chapter reviews the basic ideas of quantum mechanics. We assume that the reader has already encountered the subject at an intermediate level. Here we seek to bring together the main results needed for later chapters and to establish a consistent notation. The reader may also find our necessarily concise summary useful in systematizing previous knowledge. We shall be concerned with the results of quantum mechanics rather than the methods, which will be exemplified in later chapters. In particular, we shall assume that the states of individual molecules are known to sufficient accuracy, without enquiring how these states may have been obtained. It is a matter for quantum chemistry to establish what means have to be adopted to obtain sufficient accuracy in a given context, though we note that different properties are sensitive to different approximations.

In the following chapter we consider some quantum aspects of radiation from a largely phenomenological point of view. Then in Chapter 13 we examine in detail how the interaction of a charged particle with an electromagnetic field is incorporated into the Hamiltonian operator. Chapter 14 deals with changes of quantum state such as can be induced by an electromagnetic field, completing our coverage of principles. Chapter 15 then describes the calculation of molecular electromagnetic properties and Chapter 16 treats aspects of the theory of spectroscopy.

11.2 States

The reader will be familiar with the description of the quantum-mechanical state of a system by a wavefunction $\psi_n(x)$, where n is a label for the state and x denotes a dependence on the spatial coordinates. Instead of $\psi_n(x)$ we shall mostly write such a state as $|n\rangle$, which is known as a *ket*. This notation, due to Dirac, will simplify the form of the expressions we derive. It is also more powerful than the wavefunction representation, for example by allowing one to use the same symbol $|n\rangle$ for the state which may be described not only by $\psi_n(x)$ but also by $\phi_n(p)$, where $\phi_n(p)$ is the wavefunction in terms of the momentum; however, we shall not really exploit this power here. The complex conjugate of the wavefunction, $\psi_n(x)^*$, is represented by the *bra* $\langle n|$.

Quantum-mechanical states have the mathematical properties of vectors in a vector space (Hilbert space), and some of the nomenclature reflects this. Frequently one requires integrals of the form

$$\langle n|m\rangle = \int \psi_n^*(x)\psi_m(x)\,\mathrm{d}x, \tag{11.1}$$

which is known as an overlap integral or *scalar product*. The greater simplicity of the bra and ket notation is apparent. From eq. (11.1) we can see that the complex conjugate is

$$\langle n|m\rangle^* = \langle m|n\rangle. \tag{11.2}$$

Two states are said to be *orthogonal* if their scalar product is zero, and a state is said to be *normalized* if its scalar product with itself is unity.

Any linear combination of states, e.g. $a|n\rangle + b|m\rangle$, where a and b are scalars (i.e. numbers, which may be complex), is also a quantum-mechanical state. A set of states $\{|b_i\rangle\}$ none of which is a linear combination of any of the others is said to be *complete* if any other state $|v\rangle$ can be expanded in terms of them in the form

$$|v\rangle = \sum_i c_i |b_i\rangle, \tag{11.3}$$

where the coefficients c_i are given by $\langle b_i|v\rangle$. When the *basis* set $\{|b_i\rangle\}$ is orthogonal and normalized, or *orthonormal*, the expansion coefficients for a normalized state $|v\rangle$ satisfy

$$\sum_i |c_i|^2 = 1, \tag{11.4}$$

so that $|c_i|^2$ can be interpreted as the fraction of $|v\rangle$ which has $|b_i\rangle$ character. Expansions of this sort are particularly useful when the state $|v\rangle$ is slightly different from one of the states $|b_i\rangle$, all of which are assumed to be known (see Chapter 14).

11.3 Operators and observables

Observable quantities are associated with linear *operators* which act on one state to turn it into another, e.g.

$$|v'\rangle = \hat{A}\,|v\rangle, \tag{11.5}$$

where $\hat{A}$ is an operator as shown by the 'hat' or circumflex. Examples are the Hamiltonian operator $\hat{H}$ corresponding to the energy, the position operator $\hat{\boldsymbol{R}}$, the momentum operator $\hat{\boldsymbol{P}}$, the dipole moment operator $q\hat{\boldsymbol{R}}$, and so on. Combinations of operators yield other operators, for example the angular momentum operator $\hat{\boldsymbol{L}} = \hat{\boldsymbol{R}} \times \hat{\boldsymbol{P}}$.

The action of the operator $\hat{A}$ followed by that of the operator $\hat{B}$ is written $\hat{B}\hat{A}$, so that

$$|v''\rangle = \hat{B}\,|v'\rangle = \hat{B}\hat{A}\,|v\rangle. \tag{11.6}$$

Operators need not in general *commute*, that is, the effect of $\hat{A}\hat{B}$ may differ from that of $\hat{B}\hat{A}$. The difference operator

$$[\hat{A}, \hat{B}] \equiv \hat{A}\hat{B} - \hat{B}\hat{A} \tag{11.7}$$

is called the *commutator* of $\hat{A}$ and $\hat{B}$. In the familiar wavefunction representation, operators involve multiplication by and differentiation with respect to the coordinates: the position operator $\hat{X}$ for the x-axis is just multiplication by the coordinate x, and the corresponding component of the momentum operator $\hat{P}_x$ is $-i\hbar\partial/\partial x$, both operating on functions of x. These operators do not commute, and it may be verified using the rule for the differentiation of a product that their commutator is

$$[\hat{X}, \hat{P}_x] = i\hbar. \tag{11.8}$$

Here $\hbar$ is the Planck constant divided by 2π.

Each operator has a special set of *eigenstates* with the property that when acted upon by the operator they are affected only by being multiplied by a scalar quantity, the *eigenvalue*. If $|a_i\rangle$ is an eigenstate of $\hat{A}$, then

$$\hat{A}\,|a_i\rangle = A_i\,|a_i\rangle, \tag{11.9}$$

where A_i is the corresponding eigenvalue. The eigenvalue equation for the Hamiltonian operator is

$$\hat{H}\,|n\rangle = E_n\,|n\rangle \tag{11.10}$$

which is the time-independent Schrödinger equation, with E_n the energy of the state $|n\rangle$. The particular importance of the eigenvalues is that a measurement of the observable corresponding to the operator $\hat{A}$ must yield a number which is one of the eigenvalues A_i of $\hat{A}$. This measurement incidentally forces the system into the eigenstate $|a_i\rangle$, which is why A_i is measured. Operators which commute can be shown to have a set of states which are simultaneously eigenstates for each operator, and hence can have several eigenvalues or quantum numbers to label them, one for each operator.

Since eigenvalues are measureable quantities, they must be real numbers. This means that not all operators can correspond to observables, but only those which have real eigenvalues, called *Hermitian* operators. A Hermitian operator has other special properties. Its eigenstates form a complete orthonormal set, and so are a natural basis for calculations involving the operator in question. If eigenstates have the same eigenvalue, i.e. are *degenerate*, they are not automatically orthogonal as non-degenerate eigenstates are, but linear combinations of degenerate eigenstates can always be constructed to be orthogonal. A Hermitian operator also satisfies

$$\langle m|\hat{A}|n\rangle = \langle n|\hat{A}|m\rangle^*, \tag{11.11}$$

which is often taken as the definition of Hermiticity. Here the left-hand side means the scalar product of $\langle m|$ with the ket produced by $\hat{A}$ operating on $|n\rangle$, often referred to as a *matrix element* of $\hat{A}$ between the states $\langle m|$ and $|n\rangle$; in the usual wavefunction representation, matrix elements are integrals (being scalar products, as in eq. (11.1)), so that for example $\langle m|\hat{A}|n\rangle$ becomes $\int\psi_m^*(x)\hat{A}\psi_n(x)\,\mathrm{d}x$.

In general, when we perform a measurement of the observable associated with the operator $\hat{A}$, the system will start not in an eigenstate of $\hat{A}$ but in some other state $|v\rangle$. As already stated, the measurement must yield one of the eigenvalues A_i, but a series of measurements on a large number of identical states $|v\rangle$ will not all yield the same A_i — if they did, $|v\rangle$ would be the eigenstate $|a_i\rangle$, contrary to hypothesis. Instead the average or *expectation value* of $\hat{A}$ in the state $|v\rangle$ is obtained:

$$\langle\hat{A}\rangle \equiv \langle v|\hat{A}|v\rangle \tag{11.12}$$

$$= \sum_i |d_i|^2 A_i. \tag{11.13}$$

Here d_i is the expansion coefficient $\langle a_i|v\rangle$ as in eq. (11.3). Since $|d_i|^2$ is the fraction of $|a_i\rangle$ character in $|v\rangle$, we can argue from eq. (11.13) that it is likewise the probability of obtaining the result A_i for a measurement on the state $|v\rangle$. This measurement can also be characterized by an *uncertainty* ΔA representing the root mean square deviation of a set of measurements from the expectation value:

$$(\Delta A)^2 = \langle\hat{A}^2\rangle - \langle\hat{A}\rangle^2. \tag{11.14}$$

Only if $|v\rangle$ is an eigenstate of $\hat{A}$ is the uncertainty zero.

For two different operators $\hat{A}$ and $\hat{B}$ the product of the uncertainties cannot fall below a value determined by the commutator $[\hat{A}, \hat{B}]$; only if the operators commute can both uncertainties be zero, since then the state $|v\rangle$ can be an eigenstate of both operators simultaneously. This is the *uncertainty principle* of Heisenberg, commonly met in the particular form for the coordinate and momentum operators

$$\Delta X \Delta P_x \geqslant \tfrac{1}{2}\hbar. \tag{11.15}$$

Eigenstates of position are perfectly localized at one point, whereas eigenstates of momentum are perfectly delocalized over all points; this fundamental incompatibility of the eigenstates, implied by the non-zero commutator, means that as the state $|v\rangle$ approaches a single eigenstate of position with $\Delta X \to 0$ it also approaches a superposition of all eigenstates of momentum with $\Delta P_x \to \infty$. Only the imperfect compromise implied by the equality in eq. (11.15) can be attained.

11.4 Time dependence

Arguably the most important operator is the Hamiltonian $\hat{H}$. Not only does it correspond to the observable energy, it also determines the time development of a system. Now as far as observable quantities are concerned, what we can actually detect changing in time is the expectation value $\langle \hat{A} \rangle$ or $\langle v | \hat{A} | v \rangle$ of an operator $\hat{A}$ for a system in a state $|v\rangle$. This includes the special case when $|v\rangle$ is an eigenstate of $\hat{A}$ and we observe a single eigenvalue. There is, however, no way of determining whether the change is due to changes in the state $|v\rangle$, in the operator $\hat{A}$, or in both. We are therefore at liberty to assign the time dependence wherever is most convenient, provided that we do so consistently and without contradicting experimental observation. In the *Schrödinger picture*, all the time dependence is assigned to the states; operators other than the Hamiltonian are constant. On the other hand, in the *Heisenberg picture* all the time dependence is assigned to the operators, and states are constant. There are also intermediate pictures in which a system is described by some non-interacting and hence soluble Hamiltonian plus an interaction, when it proves convenient to have states which change in time only through the interaction instead of through the full Hamiltonian; this *interaction picture* is not used in this book.

We shall in fact use the common Schrödinger wavefunction picture, in which the time-independent Schrödinger equation (11.10) becomes

$$\hat{H}\eta_n(x) = E_n \eta_n(x), \tag{11.16}$$

where $\eta_n(x)$ is the nth eigenfunction. The time-dependent Schrödinger equation is then

$$\hat{H}\eta(x, t) = i\hbar \partial \eta(x, t)/\partial t, \tag{11.17}$$

where the wavefunction $\eta(x, t)$ depends on both position and time so that partial derivatives are required. Note that eq. (11.17) is no longer an eigenvalue equation: $\hat{H}$ determines the time dependence of any state, which could be an eigenstate of some other operator but not necessarily of $\hat{H}$ itself. Note also that $\hat{H}$ may be time-dependent in either picture, for instance if the potential energy in a system is varying.

In the case where $\hat{H}$ is independent of time, the form of the time-dependent solutions $\eta(x, t)$ of eq. (11.17) is simplified. Because $\hat{H}$ then depends only on x while the operator on the right-hand side of eq. (11.17) depends only on t, the solution $\eta(x, t)$ becomes factorizable into the product of separate functions of x and t alone:

$$\eta(x, t) = \eta(x)\, U(t) \tag{11.18}$$

Substitution in eq. (11.17) shows that $\eta(x)$ is one of the energy eigenfunctions $\eta_n(x)$, while $U(t)$ satisfies

$$i\hbar\, \mathrm{d}U(t)/\mathrm{d}t = E_n U(t) \tag{11.19}$$

with the solution

$$U(t) = U(0) \exp(-iE_n t/\hbar). \tag{11.20}$$

The constant $U(0)$ can be absorbed in the normalization, leaving

$$\eta_n(x, t) = \eta_n(x) \exp(-iE_n t/\hbar). \tag{11.21}$$

The solutions thus have an oscillatory phase factor of frequency $E_n/\hbar$, but for all times

$$|\eta_n(x, t)|^2 = |\eta_n(x)|^2. \tag{11.22}$$

As already noted, time dependence is observed experimentally in expectation values $\langle \hat{A} \rangle \equiv \langle v | \hat{A} | v \rangle$. With the help of the time-dependent Schrödinger equation and the fact that $\hat{H}$ and the operator $\hat{A}$ are Hermitian, the time derivative of $\langle \hat{A} \rangle$ can be evaluated as

$$\frac{\mathrm{d}\langle \hat{A} \rangle}{\mathrm{d}t} = \frac{1}{i\hbar} \langle \hat{A}\hat{H} - \hat{H}\hat{A} \rangle = \frac{1}{i\hbar} \langle [\hat{A}, \hat{H}] \rangle. \tag{11.23}$$

It is therefore the commutator between an operator and the Hamiltonian operator which determines the rate of change of the expectation value of the corresponding observable. If $\hat{A}$ commutes with $\hat{H}$, then $\langle \hat{A} \rangle$ is independent of time, and is said to be a *constant of the motion*. Since the Hamiltonian commutes with itself, the energy is necessarily a constant of the motion, conserved during the time development of the system.

The generalized uncertainty principle tells us that $[\hat{A}, \hat{H}]$ determines the product of the uncertainties $\Delta A \Delta H$,where we can identify ΔH with the energy uncertainty ΔE. Using this result in eq. (11.23) yields

$$t_A\, \Delta E \geqslant \tfrac{1}{2}\hbar. \tag{11.24}$$

Here the quantity t_A is the *evolution time* for the operator $\hat{A}$, defined by

$$t_A = \Delta A / |\mathrm{d}\langle \hat{A} \rangle / \mathrm{d}t|. \tag{11.25}$$

This corresponds to the time required for the expectation value to change by the uncertainty and hence to become distinguishable or resolvable from its previous value — see Fig. 11.1.

Equation (11.24) is referred to as the energy-time uncertainty principle, though we see that what is involved is not the parameter time itself. The

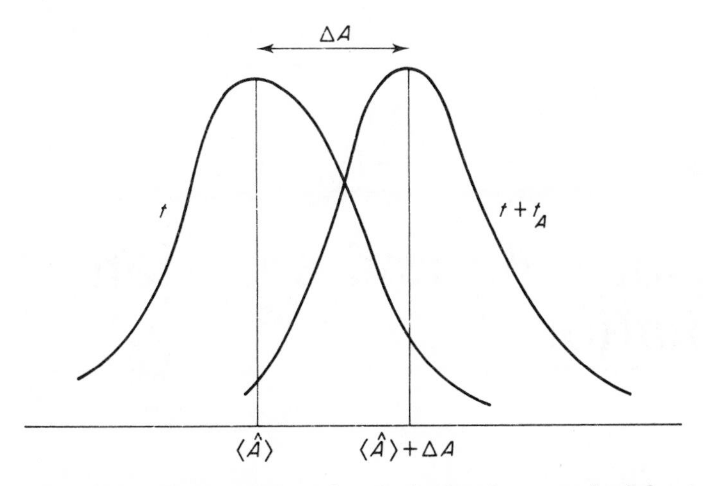

Fig. 11.1 The evolution time t_A is the time required for the distribution of measured values of the observable corresponding to the operator $\hat{A}$ to change from that at time t, with expectation value $\langle \hat{A} \rangle$ and uncertainty ΔA, to one with expectation value $\langle \hat{A} \rangle + \Delta A$. The two distributions are then resolvable according to the usual criteria for optical instruments

consequence of this expression is that states of well-defined energy, with a very narrow energy spread ΔE, must have correspondingly long evolution times or lifetimes, while broad diffuse energy states decay rapidly. Conversely, short lived states have poorly-defined energies. This result has spectroscopic consequences, since in Chapter 12 we shall see that all excited states have finite lifetimes because of spontaneous emission. Excited states therefore have finite energy widths, so that spectral lines have natural widths. A quantitative relationship between the lifetime and the natural linewidth is derived in Chapter 14. However, in practice only the most careful and precise measurements are able to approach the natural linewidth in resolution. For example, for an electronic state with $t_A \approx 1$ ns, ΔE is of the order of 10^{-7} eV, requiring a resolution of the order of 10^{-3} cm^{-1} in wavenumber.

Further reading

J. L. Martin, *Basic Quantum Mechanics*, Clarendon Press, Oxford, 1981.

P. Landshoff and A. Metherell, *Simple Quantum Physics*, Cambridge University Press, 1979.

D. T. Gillespie, *A Quantum Mechanics Primer*, International Textbook Company, Scranton PA, 1970.

J. P. Lowe, *Quantum Chemistry*, Academic Press, New York, 1968.

I. N. Levine, *Quantum Chemistry*, 3rd edn, Allyn and Bacon, Boston, 1983.

Chapter 12

Absorption and emission of radiation

12.1 Photons

In Chapter 6 we have treated some aspects of electromagnetic radiation: electromagnetic waves themselves, and the emission of electromagnetic waves by a classical oscillating dipole. In the present chapter and those which follow we treat aspects of the quantum mechanics of the absorption and emission process. In later chapters, our approach is often semi-classical; that is, we treat the radiation classically and the molecular states quantum-mechanically. Those features of the quantum statistical mechanics of radiation which we need to consider are collected in this chapter.

In fact, it was the properties of light which forced the recognition that there must be such things as quanta. Classical statistical mechanics was used by Rayleigh and Jeans to calculate the energy density as a function of wavelength for radiation in thermal equilibrium at a given temperature. This function is known as the *black-body distribution,* since a black (i.e. perfectly absorbing) body is necessary to ensure the thermal equilibrium. The calculation agreed with experiment at long wavelengths but predicted an energy density which diverged at short wavelengths where experiment showed it actually fell to zero—the 'ultraviolet catastrophe' illustrated in Fig. 12.1.

Measurements of the photoelectric effect were also inexplicable classically. It was found that no electrons were emitted from a metal plate if it was illuminated with light of a wavelength greater than some value characteristic of the metal, no matter what the intensity of the light, i.e. its energy. For smaller wavelengths electrons were emitted whatever the light intensity.

Planck and Einstein showed that these classically puzzling results could be explained with the additional hypothesis that the energy of electromagnetic radiation was obtainable only in discrete amounts or *quanta.* For light of frequency ν the energy is

$$E = h\nu = hc/\lambda, \tag{12.1}$$

where c is the speed of light and λ is the wavelength. With this hypothesis, the

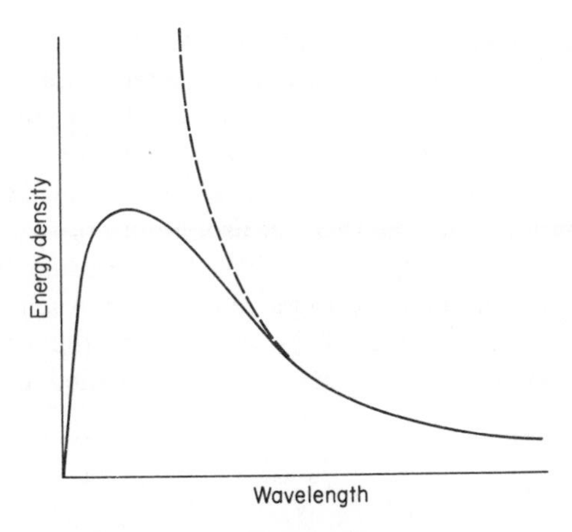

Fig. 12.1 Black-body energy distribution, showing the classical prediction of a divergence at short wavelengths (broken line—the 'ultraviolet catastrophe') and the experimental distribution (solid line) which the quantum-mechanical theory predicts correctly

ultraviolet catastrophe disappears because at short wavelengths each quantum has a large and hence thermally improbable energy; the proper black-body distribution is derived in the following section. The photoelectric effect is explained if an electron can be emitted only by absorbing a quantum of high enough energy, or small enough wavelength. A high intensity composed of many low-energy quanta will therefore not cause emission, while a low intensity composed of a few high-energy quanta will.

The quanta of electromagnetic radiation are called *photons*, the suffix 'on' indicating their particle-like nature. They can often be treated as particles of energy $h\nu$ and momentum h/λ (for example, in treating the Compton effect for scattering of electrons by light). They have spin $s = 1$, but because they have zero rest mass they have not the usual $2s + 1 = 3$ states, but only two. Classically these can be regarded as equivalent to right and left-hand circular polarizations. Since total angular momentum is conserved, a molecule which emits or absorbs a photon must have its state of angular momentum changed by a corresponding amount. This fact accounts for the existence of *selection rules* determining the states between which molecules may make transitions involving photons. Selection rules are treated from a different viewpoint in Chapter 16.

Particles of integer spin obey what are known as Bose–Einstein statistics, and are often called *bosons.* Wavefunctions for such particles must be symmetric under interchange of identical particles, for example by relabelling them, and this allows any number of the particles to occupy the same quantum state, excluding the spin part. (Particles of half-integer spin, most notably electrons, obey Fermi–Dirac statistics, and are called *fermions*. Only a limited number of such particles can occupy the same state, depending on the spin; electrons are limited to two per state, as the Pauli exclusion principle indicates.) If we assume that the thermal probability of occupation of a state containing n photons of energy $h\nu$ is given by the usual Boltzmann factor $\exp(-nh\nu/kT)$, the thermal average number of such photons at the temperature T is readily evaluated. It is given by the sum of all photon numbers n weighted by the Boltzmann factor, divided by the sum of the weights:

$$N(\nu, T) = \sum_{n=0}^{\infty} n\, \mathrm{e}^{-nh\nu/kT} \Big/ \sum_{m=0}^{\infty} \mathrm{e}^{-mh\nu/kT}. \tag{12.2}$$

If we set $x = \mathrm{e}^{-h\nu/kT}$ it can be seen that the sums required are the elementary ones

$$\sum_{n=0}^{\infty} x^n = 1/(1-x), \tag{12.3}$$

$$\sum_{n=0}^{\infty} nx^n = x/(1-x)^2. \tag{12.4}$$

Hence we obtain

$$N(\nu, T) = \mathrm{e}^{-h\nu/kT}/(1-\mathrm{e}^{-h\nu/kT}) \tag{12.5}$$

$$= 1/(\mathrm{e}^{h\nu/kT} - 1) \tag{12.6}$$

It is often convenient to use instead of ν the radian frequency $\omega = 2\pi\nu$. Correspondingly we introduce the modified Planck constant $\hbar = h/2\pi$, in terms of which the photon energy becomes $\hbar\omega$.

12.2 Black-body distribution

As already noted, the black-body distribution gives the energy density for radiation in thermal equilibrium. From the preceding section we know the photon energy and the thermal equilibrium number of photons in a given state. All that remains is to calculate the number of possible photon states. We shall in fact calculate the number of such states per unit frequency interval, to give the energy density per unit frequency interval.

For simplicity we suppose that the electromagnetic wave in question is confined in a cube of side L. The shape of the container does not appear in the final result, which can be derived more generally. The space part of the wave can be written as

$$\exp[(2\pi i\nu/c)(\boldsymbol{r}\cdot\boldsymbol{k})] = \exp[(2\pi i\nu/c)(xk_x + yk_y + zk_z)], \tag{12.7}$$

where $\boldsymbol{k}$ is a unit vector in the direction of propagation. If the wave is to be reflected at the walls without suffering destructive interference it must satisfy the conditions for a standing wave. These require the same amplitude at $x = 0$ and L, $y = 0$ and L, and $z = 0$ and L. Thus we need, for example,

$$\exp[(2\pi i\nu/c)Lk_x] = 1 \tag{12.8}$$

Since $\exp(2\pi i n) = 1$ for integer n, eq. (12.8) implies

$$\nu L k_x/c = n_x, \tag{12.9}$$

where n_x is some integer giving the number of half-waves fitting across the cube in the x direction.

Thus each allowed wave or photon state is represented by a triple of integers (n_x, n_y, n_z) which satisfy

$$n^2 = n_x^2 + n_y^2 + n_z^2 = (\nu L/c)^2(k_x^2 + k_y^2 + k_z^2) \tag{12.10}$$

$$= (\nu L/c)^2, \tag{12.11}$$

since $\boldsymbol{k}$ is a unit vector. The number of allowed waves N with a frequency up to ν is the number of points inside a sphere of radius n given by eq. (12.11), namely

$$N = (4\pi/3)(\nu L/c)^3. \tag{12.12}$$

The number of such states per unit frequency interval is then

$$\mathrm{d}N/\mathrm{d}\nu = 4\pi\nu^2 L^3/c^3. \tag{12.13}$$

We recognize L^3 as the volume in which the photons are confined, and recall that each wave can have two different polarizations. Then the total number of photon states per unit volume per unit frequency interval is

$$n(\nu) = 2(\mathrm{d}N/\mathrm{d}\nu)/L^3 \tag{12.14}$$

$$= 8\pi\nu^2/c^3. \tag{12.15}$$

This may be referred to as simply the photon density of states, with 'density' doing service to mean both 'per unit volume' and 'per unit frequency'. Precise specification is preferable to avoid confusion, but a check on dimensions is often helpful: in eq. (12.15) the dimensions are easily checked as 1/(volume × frequency). One also meets densities of states defined analogously to (12.14) but with $\mathrm{d}N/\mathrm{d}\nu$ replaced by $\mathrm{d}N/d\omega$, where ω is the radian frequency $2\pi\nu$ as usual, by $\mathrm{d}N/\mathrm{d}\sigma$, where σ is the wavenumber or inverse wavelength ν/c, or by $\mathrm{d}N/\mathrm{d}\lambda$, where λ is the wavelength (see Examples, p. 156). With the number density given by eq. (12.15), the energy density is just $h\nu$ times $n(\nu)$. The thermal average energy density per unit volume per unit frequency interval is then

$$\varrho(\nu, T) = h\nu n(\nu)N(\nu, T), \tag{12.16}$$

where $N(\nu, T)$ is obtained from eq. (12.6), giving

$$\varrho(\nu, T) = \frac{8\pi h\nu^3/c^3}{e^{h\nu/kT} - 1}. \tag{12.17}$$

This is the black-body distribution law. It agrees exactly with the experimental curve shown in Fig. 12.1 (the solid line). The classical equivalent is obtained by letting $h \to 0$, since the appearance of factors of h is characteristic of quantum-mechanical results, or equivalently by letting $T \to \infty$. The exponential in the denominator of eq. (12.17) can then be approximated by $1 + h\nu/kT$, to yield

$$\varrho_{\text{class}}(\nu, T) = 8\pi kT\nu^2/c^3. \tag{12.18}$$

which is independent of h as required and corresponds to the broken line in Fig. 12.1. This form is approached by the exact expression at low frequencies and high temperatures. It is the exponential in eq. (12.17) which suppresses the ultraviolet catastrophe by increasing even faster than ν^3 as $\nu \to \infty$, so making $\varrho(\nu, T) \to 0$: the energy quantum becomes so large that the thermal average number of photons falls off faster than the number of photon states increases.

12.3 Einstein coefficients

The black-body distribution just derived treats the thermal equilibrium distribution of photon energies. This is a statistical property: the black body with which the radiation is in equilibrium will be continually absorbing and emitting photons, but the net rates of absorption and emission will be equal. We therefore now move on to consider the rates of absorption and emission in a molecular system and their relationship to the dynamic equilibrium already considered. This phenomenological treatment was introduced by Einstein.

Suppose that a molecular system has an upper state u and a lower state l, with N_u molecules in the upper state and N_l in the lower. Elementary kinetic arguments indicate that the molecules in the lower state will absorb photons at a rate proportional to both the number of such molecules and the density of photons. This *stimulated absorption* is described by the kinetic equation

$$-\mathrm{d}N_l/\mathrm{d}t = N_l B_{lu}(\nu)\varrho(\nu), \tag{12.19}$$

where T is omitted in the density of photon states $\varrho(\nu)$ because there need no longer be thermal equilibrium. The coefficient $B_{lu}(\nu)$ is the *Einstein coefficient of stimulated absorption.* As indicated, B_{lu} is expected to depend on the frequency, but it is not expected to depend on temperature. Similarly, the molecules in the upper state will emit photons at a rate proportional to the number of molecules and the photon density. *Stimulated emission* is described by the kinetic equation

$$-\mathrm{d}N_u/\mathrm{d}t = N_u B_{ul}(\nu)\varrho(\nu), \tag{12.20}$$

where B_{ul} is the *Einstein coefficient of stimulated emission*. (The need for photons to stimulate photon emission may not be obvious, but is analogous to the use of electron bombardment to produce positive ions in the mass spectrometer, and is shown to arise naturally when a periodic perturbation causes a change of quantum state—see Chapter 14.)

In the steady state, the rate at which molecules are lost from the lower state to the upper by absorption must equal the rate at which molecules are gained in the lower state from the upper by emission. The numbers of molecules in the two states must also be related by the Boltzmann factors:

$$N_u = N_l \exp(-h\nu/kT), \tag{12.21}$$

since the photon energy $h\nu$ is clearly the difference in energy between the upper and lower states. If we equate the right-hand sides of eqs (12.19) and (12.20) and use eq. (12.21), we find that B_{lu} and B_{ul} must be related by the Boltzmann factor—but this is contrary to the assumption that each is independent of temperature. From such considerations Einstein deduced that there must be a third process, emission at a rate proportional to the number of molecules in the upper state but independent of the photon density. This *spontaneous emission* is described by the kinetic equation

$$-\mathrm{d}N_u/\mathrm{d}t = N_u A_{ul}(\nu), \tag{12.22}$$

where A_{ul} is the *Einstein coefficient of spontaneous emission*.

If we now examine the steady state, we find that the Einstein coefficients can be independent of temperature but are not independent of one another. Equating the rate of absorption from eq. (12.19) to the net rate of emission from eqs (12.20) and (12.22), we obtain

$$N_l B_{lu}(\nu)\varrho(\nu, T) = N_u[B_{ul}(\nu)\varrho(\nu, T) + A_{ul}(\nu)]. \tag{12.23}$$

With eq. (12.21), this can be rearranged to express $\varrho(\nu, T)$ as

$$\varrho(\nu, T) = \frac{A_{ul}(\nu)}{B_{lu}(\nu)\mathrm{e}^{h\nu/kT} - B_{ul}(\nu)} \tag{12.24}$$

This expression for $\varrho(\nu, T)$ must of course equal that given by eq. (12.17), as each refers to the thermal equilibrium steady state. In order to obtain the denominator $\mathrm{e}^{h\nu/kT} - 1$, we must have

$$B_{lu}(\nu) = B_{ul}(\nu), \tag{12.25}$$

i.e. the Einstein coefficients of stimulated absorption and emission are equal. The whole expression for $\varrho(\nu, T)$ then requires

$$A_{ul}(\nu) = (8\pi h\nu^3/c^3)B_{lu}(\nu) \tag{12.26}$$

Thus a knowledge of one Einstein coefficient suffices to determine the others.

Note that coefficients of stimulated absorption and emission could be defined in terms of the density of photon states per unit radian frequency ω or per unit wavenumber σ. This would change ϱ, and B would then have to change correspondingly to describe the same physical rate of absorption or emission of photons. However, the coefficient of spontaneous emission is independent of the photon density and so does not change with the choice of ϱ.

Although the stimulated coefficients are equal, the rates they describe are not, because they depend on N_l and N_u, which must differ in thermal equilibrium by eq. (12.21). Spontaneous emission is necessary to achieve the balanced steady state. The total rate of absorption can be written as

$$-\mathrm{d}N_l/\mathrm{d}t = N_l B_{lu}(\nu) n(\nu) h\nu N(\nu, T) \tag{12.27}$$

and the total rate of emission as

$$-\mathrm{d}N_u/\mathrm{d}t = N_u B_{ul}(\nu)[n(\nu)h\nu N(\nu, T) + 8\pi h\nu^3/c^3], \tag{12.28}$$

where eq. (12.16) has been used for $\varrho(\nu, T)$ and eq. (12.26) has been used to eliminate $A_{ul}(\nu)$. But from eq. (12.15), $n(\nu)h\nu$ is equal to $8\pi h\nu^3/c^3$, so that we can write

$$-\mathrm{d}N_u/\mathrm{d}t = N_u B_{ul}(\nu) n(\nu) h\nu [N(\nu, T) + 1]. \tag{12.29}$$

The rate of absorption is proportional to the thermal average number of photons $N(\nu, T)$, but the rate of emission is proportional to $N(\nu, T) + 1$, which is clearly non-zero even when there are no photons present. This result was shown by Dirac to be a consequence of quantizing the radiation field.

12.4 Absorption intensity

Since the Einstein coefficients determine the rate of absorption of photons, they also determine the absorption intensity. They can therefore be related to the conventional spectroscopic measures of intensity. If an intensity I is transmitted through a length x of an absorbing medium containing a concentration C of absorbing species when the incident intensity is I_0, then the *absorbance* or optical density D is given by

$$D = \log_{10}(I_0/I) = \epsilon C x, \tag{12.30}$$

where ϵ is the *molar absorption* (or extinction) *coefficient*. This should not be confused with the dielectric permittivity given the same symbol, which does not occur in the present chapter. Correspondingly we have

$$I = I_0 10^{-\epsilon C x}, \tag{12.31}$$

which is the *Beer–Lambert law*. In this section ϵ will be related to the Einstein coefficient B_{lu}. For this purpose it is convenient to use yet another form of the law, giving the fractional change in intensity for light passing through an element of the absorbing medium of length dx:

$$\mathrm{d}I/I = -\epsilon C\,\mathrm{d}x \ln 10 \tag{12.32}$$

where the factor ln 10 arises from converting to an exponential in order to perform the differentiation of eq. (12.31). The dimensions of ϵ are seen to be 1/(concentration × length), so that $m^2\ mol^{-1}$ would be appropriate units. (The unit *dark* is also met, in which the length is taken as 1 cm and the concentration as 1 $mmol\ dm^{-3}$, so that 1 dark is 100 $m^2\ mol^{-1}$, but this non-SI unit is not recommended.)

In practice, intensities cannot be measured at a single frequency ν but only over a range $d\nu$ centred on ν. There are two reasons for this. First, excited states necessarily have finite lifetimes because of spontaneous emission, which means that they also have an uncertainty in energy, as discussed in the last chapter. The transition between two states cannot then be infinitely sharp but must have a width related to the inverse lifetime (see also Chapter 14). Secondly, spectroscopic instruments have a finite resolution, for instance because entrance slits cannot be infinitesimally narrow as this would prevent all light from reaching the detector. We are therefore normally concerned with the change in intensity integrated over a band of width $\Delta\nu$ (which may of course be small), and thus with the integrated molar absorption coefficient

$$A = \ln 10 \int_{\nu-\frac{1}{2}\Delta\nu}^{\nu+\frac{1}{2}\Delta\nu} \epsilon(\nu')\, d\nu'. \tag{12.33}$$

Now the intensity is the power density, i.e. the energy per unit time falling on unit area. It decreases as molecules absorb photons, so removing energy. We can write the intensity change as the light passes through a length dx of the absorbing medium in the form

$$dI = (dN_l/dt)h\nu/S \tag{12.34}$$

where S is the area on which the light is incident. For infrared and ultraviolet absorption, unless the incident intensity is very high, the overwhelming majority of molecules will remain in the lower state, so that the net rate of absorption will be proportional to N_l as usual, but this in turn will be essentially the total number of molecules N. Note, however, that one important situation where this assumption would be wrong is in a laser, where the mechanism depends on having a population inversion $N_u > N_l$ between two levels rather than $N_u \ll N_l$; but lasers and non-linear effects are not treated here. The treatment of intensity is also somewhat different in microwave and radiofrequency spectroscopy, where at ordinary temperatures the energy difference between the upper and lower levels is so small compared with kT that their thermal equilibrium populations are nearly equal. In such circumstances it is possible at modest irradiation power to equalize the populations and *saturate* the absorption.

The Einstein coefficients were introduced with the tacit assumption of a sharp transition between the upper and lower levels at the frequency ν. If the transition is broadened for the reasons explained above, the coefficients describe the *total* rate over the bandwidth $\Delta\nu$. Thus in the region $d\nu$ round ν we have

$$-dN_l/dt = B_{lu}N\varrho(\nu)\, d\nu/\Delta\nu \tag{12.35}$$

where $\varrho(\nu)$ is the photon density as before, with T omitted because there is no longer thermal equilibrium in the system as it is absorbing photons. In time dt light travels a distance $c\ dt$, where c is the speed of light (essentially that *in vacuo* for a dilute gaseous medium), so that in that time interval an area S is hit by the photons in a volume $Sc\ dt$. The intensity is therefore

$$I = c\varrho(\nu)\ d\nu \tag{12.36}$$

since the areas S and time interval dt cancel out, so that

$$-dN_l/dt = B_{lu}NI/c\Delta\nu. \tag{12.37}$$

Substitution of eq. (12.37) in (12.34) now gives

$$dI/I = -B_{lu}Nh\nu/cS\Delta\nu. \tag{12.38}$$

The concentration (amount per unit volume) is

$$C = N/LV = N/LS\ dx, \tag{12.39}$$

where L is the Avogadro constant and V is the volume containing the N molecules. Thus we obtain

$$dI/I = -(B_{lu}Lh\nu/c\Delta\nu)C\ dx \tag{12.40}$$

whence from eq. (12.32) we may identify the molar absorption coefficient as

$$\epsilon \ln 10 = B_{lu}Lh\nu/c\Delta\nu. \tag{12.41}$$

Finally the integrated molar absorption coefficient is

$$A = B_{lu}Lh\nu/c \tag{12.42}$$

where ν is a mean absorption frequency for the band.

The absorption coefficient may also be integrated with respect to wavenumber $\sigma = \nu/c$ instead of frequency, giving $A' = B_{lu}Lh\sigma/c$, or $B'_{lu}Lh\sigma$ if a new Einstein coefficient B'_{lu} is defined with respect to the photon density $\varrho(\sigma)$ instead of $\varrho(\nu)$ as discussed in the previous section. The dimensions of A' are those of ϵ times 1/length, so that suitable units would be m mol^{-1}. In all these measures of intensity it has been assumed that the band is narrow enough for the Einstein coefficient not to vary significantly; otherwise a further integration is necessary.

Examples

12.1 Starting from eq. (12.12), calculate $n(\omega)$, the number of photon states per unit volume per unit radian frequency $\omega = 2\pi\nu$. Hence calculate the black-body energy distribution $\varrho(\omega,T)$. How must the Einstein coefficient $B_{lu}(\omega)$ be related to $B_{lu}(\nu)$ in order to leave the left-hand side of eq. (12.19) unchanged?

12.2 Repeat Example 12.1 in terms of the wavenumber $\sigma = \nu/c$, calculating $n(\sigma)$, $\varrho(\sigma,T)$ and $B_{lu}(\sigma)$.

12.3 Repeat Example 12.1 in terms of the wavelength $\lambda = c/\nu$ to obtain

$$\varrho(\lambda,T) = \frac{8\pi hc/\lambda^5}{e^{hc\lambda/kT} - 1}.$$

12.4 Derive an expression for the wavelength λ_{max} at which $\varrho(\lambda, T)$ is a maximum.. Show that $\lambda_{max}T$ is a constant, which is approximately $hc/5k$. This is *Wien's displacement law.*

12.5 Calculate the total energy density per unit volume u by integrating $\varrho(\nu, T)$ over all ν from 0 to ∞. The result requires the integral

$$\int_0^\infty \frac{y^3\,\mathrm{d}y}{\mathrm{e}^y - 1} = \frac{\pi^4}{15}.$$

Hence deduce that the radiant emittance $R = cu/4$ of a black body is proportional to T^4 and evaluate the constant of proportionality. This is the *Stefan–Boltzmann law.*

12.6 From the Einstein rate equations for the absorption and emission of photons, derive an expression for the equilibrium value of the fractional population difference $\Delta N/N$, where $\Delta N = N_l - N_u$ and $N = N_l + N_u$. Substituting for the Einstein coefficients and the photon energy density, show that

$$\Delta N/N = 1/[2N(\nu, T) + 1].$$

Verify that this result is consistent with the Boltzmann distribution.

12.7 Verify that the integrated absorption A' is given by $B_{lu}(\sigma)Lh\sigma$ with $B_{lu}(\sigma)$ from Example 12.2.

Further reading

D. J. E. Ingram, *Radiation and Quantum Physics* (Clarendon Press, Oxford, 1973).

P. W. Atkins, *Molecular Quantum Mechanics,* 2nd edn (Clarendon Press, Oxford, 1983).

W. P. Healy, *Non-Relativistic Quantum Electrodynamics* (Academic Press, London, 1982).

J. I. Steinfeld, *Molecules and Radiation* (Harper and Row, New York, 1974).

P. W. Milonni, Why spontaneous emission? *Am. J. Phys.,* **52** (1984) 340.

Chapter 13

The Hamiltonian in electromagnetic fields

In Chapter 11 we summarized basic principles of quantum mechanics, with a bias towards applications in the rest of Part C. The Hamiltonian operator corresponds to the total energy of a system and as we saw, also determines its time evolution. It is therefore important to consider in detail the form this operator takes for a charged particle in an electromagnetic field described in the usual way by a scalar potential V and a vector potential $\boldsymbol{A}$.

13.1 Generalized momentum

The Hamiltonian operator is obtained from the classical *Hamiltonian function* H by replacing the generalized coordinates and momenta in H by the corresponding operators. Often this is straightforward: H can be written as the sum of the kinetic energy T and the potential energy Φ, and in simple cases T depends only on the momenta and Φ only on the coordinates. (We use Φ for the potential energy rather than the more usual V to avoid confusion with the scalar potential.) For a particle in a purely electrostatic potential this is the case, and the potential energy qV for a charge q often appears in Hamiltonian operators without comment. However, the vector potential affects the momentum.

To deal with this problem, it is necessary to make an excursion into classical mechanics. The Hamiltonian function is defined in terms of the generalized momentum $\boldsymbol{p}$, to be defined below, as

$$H = \boldsymbol{p} \cdot \boldsymbol{v} - L \tag{13.1}$$

where $\boldsymbol{v}$ is the particle velocity and L is the *Lagrangian function*. In using this equation, one must express $\boldsymbol{v}$ in terms of $\boldsymbol{p}$. The Lagrangian has to be constructed so that the equations of motion of the system take the form

$$\frac{\mathrm{d}}{\mathrm{d}t}\left(\frac{\partial L}{\partial v_\alpha}\right) = \frac{\partial L}{\partial r_\alpha}, \tag{13.2}$$

whre v_α and r_α are components of the velocity $\boldsymbol{v}$ and the position $\boldsymbol{r}$. Then the *generalized momentum* has components defined by

$$p_\alpha = \partial L/\partial v_\alpha \tag{13.3}$$

It is not in general equal to mv, which may be called the *kinetic* momentum $\boldsymbol{\pi}$:

$$\boldsymbol{\pi} = m\boldsymbol{v}. \tag{13.4}$$

For a particle of charge q in an electric field $\boldsymbol{E}$ and a magnetic induction $\boldsymbol{B}$, we know that the motion is determined by the Lorentz force (Section 3.9)

$$\boldsymbol{F} = q[\boldsymbol{E} + (\boldsymbol{v} \times \boldsymbol{B})] = \mathrm{d}\boldsymbol{\pi}/\mathrm{d}t. \tag{13.5}$$

We now introduce the potentials and rearrange this equation into a form like eq. (13.2). We have

$$\boldsymbol{E} = -\nabla V - \partial \boldsymbol{A}/\partial t, \tag{13.6}$$

$$\boldsymbol{B} = \nabla \times \boldsymbol{A}, \tag{13.7}$$

with the result

$$m\left(\frac{\mathrm{d}\boldsymbol{v}}{\mathrm{d}t}\right) = q\left[-\nabla V - \frac{\partial \boldsymbol{A}}{\partial t} + \boldsymbol{v} \times (\nabla \times \boldsymbol{A})\right], \tag{13.8}$$

where by a vector identity for Cartesian components α and β

$$[\boldsymbol{v} \times (\nabla \times \boldsymbol{A})]_\alpha = \sum_\beta v_\beta \left(\frac{\partial A_\beta}{\partial r_\alpha} - \frac{\partial A_\alpha}{\partial r_\beta}\right). \tag{13.9}$$

Now the total time derivative of $\boldsymbol{A}$ is

$$\frac{\mathrm{d}\boldsymbol{A}}{\mathrm{d}t} = \frac{\partial \boldsymbol{A}}{\partial t} + \sum_\beta \frac{\partial \boldsymbol{A}}{\partial r_\beta}\frac{\mathrm{d}r_\beta}{\mathrm{d}t}. \tag{13.10}$$

Recognizing $\mathrm{d}r_\beta/\mathrm{d}t$ as the velocity v_β, we can extract $\mathrm{d}\boldsymbol{A}/\mathrm{d}t$ from the right-hand side of eq. (13.8) and transfer it to the left-hand side with the other time derivative, to obtain

$$\frac{\mathrm{d}}{\mathrm{d}t}(m\boldsymbol{v} + q\boldsymbol{A})_\alpha = q(-\nabla_\alpha V + \sum_\beta v_\beta \partial A_\beta/\partial r_\alpha) \tag{13.11}$$

This equation is now of the form (13.2) if

$$\partial L/\partial v_\alpha = mv_\alpha + qA_\alpha \tag{13.12}$$

$$\partial L/\partial r_\alpha = -q\partial V/\partial r_\alpha + q\partial(\boldsymbol{v}\cdot\boldsymbol{A})/\partial r_\alpha \tag{13.13}$$

We can integrate these partial differential equations, using the fact that the potentials are independent of $\boldsymbol{v}$, to find that the Lagrangian must satisfy the two equations

$$L = \tfrac{1}{2}mv^2 + q\boldsymbol{v}\cdot\boldsymbol{A} + f(\boldsymbol{r}), \tag{13.14}$$

$$L = -qV + q\boldsymbol{v}\cdot\boldsymbol{A} + g(\boldsymbol{v}), \tag{13.15}$$

where $f(\boldsymbol{r})$ and $g(\boldsymbol{v})$ are the complementary functions of integration. These equations are compatible if $f(\boldsymbol{r}) = -qV(\boldsymbol{r})$ and $g(\boldsymbol{v}) = \frac{1}{2}mv^2$, so that we may take the Lagrangian as

$$L = \tfrac{1}{2}mv^2 + q\boldsymbol{v}\cdot\boldsymbol{A} - qV. \tag{13.16}$$

From this Lagrangian we find that the components of the generalized momentum are given according to eq. (13.3) by

$$p_\alpha = mv_\alpha + qA_\alpha. \tag{13.17}$$

The additional velocity-dependent term in the Lagrangian in a vector potential thus gives rise to an extra term in the generalized momentum. The kinetic momentum is obtained from the generalized momentum and eq. (13.4) as

$$\pi_\alpha = mv_\alpha = p_\alpha - qA_\alpha. \tag{13.18}$$

13.2 Classical Hamiltonian

The classical Hamiltonian can now be found from eq. (13.1). With eqs (13.16) and (13.17) we find

$$H = mv^2 + q\boldsymbol{v}\cdot\boldsymbol{A} - \tfrac{1}{2}mv^2 - q\boldsymbol{v}\cdot\boldsymbol{A} + qV \tag{13.19}$$

$$= \tfrac{1}{2}mv^2 + qV. \tag{13.20}$$

This takes the usual form of the kinetic energy plus a potential energy. However, H must be expressed as a function of $\boldsymbol{p}$ rather than $\boldsymbol{v}$ or $\boldsymbol{\pi}$. From eq. (13.18) we obtain finally

$$H = (\boldsymbol{p} - q\boldsymbol{A})^2/2m + qV. \tag{13.21}$$

This derivation has assumed that the only force has been the Lorentz force. Any other force enters as the negative gradient of a potential Φ, which is carried through the algebra like the electrostatic force $-q\nabla V$, leading to an extra term Φ in eq. (13.21). Thus the Hamiltonian in an electromagnetic field is changed from that in zero field by replacing $\boldsymbol{p}$ by $\boldsymbol{p} - q\boldsymbol{A}$ and adding a term qV.

The Hamiltonian can also be derived rather directly from the relativistic considerations of Chapter 7. The relativistically invariant generalized four-momentum is

$$\mathsf{p} = (\boldsymbol{p}, iE/c), \tag{13.22}$$

where E is the total energy to which H corresponds. The equivalent kinetic four-momentum $\boldsymbol{\pi}$ must arise in a similar way with the fourth component iT/c, where T is the kinetic energy as before. Given that the potential energy in the electromagnetic field is qV, we have $T = E - qV$. The fourth component of $\boldsymbol{\pi}$ is therefore $iE/c - q(iV/c)$, where we recognize iE/c as the fourth component of **p** and iV/c as the fourth component of the four-vector potential **A**. This enables us to construct $\boldsymbol{\pi}$ at once as

$$\boldsymbol{\pi} = (p - qA,\ iT/c), \tag{13.23}$$

so that the ordinary kinetic momentum is $p - qA$ as before.

The Hamiltonian can now be derived as before, or else we can use the fact that the total energy is related to the momentum by the Klein–Gordon equation

$$E^2 = p^2c^2 + m_0^2c^4, \tag{13.24}$$

so that correspondingly the kinetic energy is given by

$$T^2 = \pi^2c^2 + m_0^2c^4, \tag{13.25}$$

Hence we find

$$(E - qV)^2 = m_0{}^2c^4 + (p - qA)^2c^2 \tag{13.26}$$

which leads to

$$E = qV + m_0c^2[1 + (p - qA)^2/m_0^2c^2]^{1/2}. \tag{13.27}$$

In the non-relativistic limit, $(p - qA)/m_0c \ll 1$, and we can expand the square root to first order and neglect higher terms. Then apart from the constant rest energy m_0c^2 we obtain

$$E = (p - qA)^2/2m_0 + qV, \tag{13.28}$$

which agrees with eq. (13.21) since at low velocities the mass m is equal to the rest mass m_0.

13.3 Quantization

The quantum-mechanical Hamiltonian operator $\hat{H}$, where the circumflex distinguishes the operator from the Hamiltonian function, is obtained from the latter (expressed as a function of the generalized momentum $\boldsymbol{p}$ and the position $\boldsymbol{r}$) by simply replacing $\boldsymbol{p}$ and $\boldsymbol{r}$ by the corresponding operators $\hat{\boldsymbol{P}}$ and $\hat{\boldsymbol{R}}$. Thus

$$\hat{H} = [\hat{\boldsymbol{P}} - qA(\hat{\boldsymbol{R}})]^2/2m + qV(\hat{\boldsymbol{R}}). \tag{13.29}$$

For a set of charges, each charge contributes a term like this to $\hat{H}$. Once again, any other potential energy has to be added to $\hat{H}$.

As we saw in Chapter 11, for purposes of computation some concrete representation is required. In the Schrödinger coordinate representation, $\hat{\boldsymbol{R}}$ is represented by multiplication by the coordinate $\boldsymbol{R}$, while $\hat{\boldsymbol{P}}$ is represented by

$-i\hbar\nabla$. Then the Hamiltonian becomes

$$\hat{H} = [i\hbar\nabla + q\boldsymbol{A}(\boldsymbol{R})]^2/2m + qV(\boldsymbol{R}) \tag{13.30}$$

$$= -(\hbar^2/2m)\nabla^2 + (i\hbar q/2m)(\boldsymbol{A}\cdot\nabla + \nabla\cdot\boldsymbol{A}) + (q^2/2m)A^2 + qV. \tag{13.31}$$

Recall that the Hamiltonian in this representation operates on the wave-function ψ, which is a scalar function of $\boldsymbol{R}$. Then the term in $\nabla\cdot\boldsymbol{A}$ yields

$$\nabla\cdot\boldsymbol{A}\psi(\boldsymbol{R}) = \psi(\boldsymbol{R})(\nabla\cdot\boldsymbol{A}) + \boldsymbol{A}\cdot\nabla\psi(\boldsymbol{R}), \tag{13.32}$$

by the usual rule for differentiation of a product. Hence

$$\hat{H} = -\frac{\hbar^2}{2m}\nabla^2 + \frac{i\hbar q}{m}\boldsymbol{A}\cdot\nabla + \frac{i\hbar q}{2m}\operatorname{div}\boldsymbol{A} + \frac{q^2A^2}{2m} + qV, \tag{13.33}$$

where we have written div $\boldsymbol{A}$ instead of $(\nabla\cdot\boldsymbol{A})$ in this equation to emphasize that the gradient operator ∇ does not operate on $\psi(\boldsymbol{R})$. For atomic and molecular calculations it is often convenient to work in the *Coulomb gauge* div $\boldsymbol{A} = 0$, as in Chapter 3. In this gauge we obtain the final result

$$\hat{H} = -\frac{\hbar^2}{2m}\nabla^2 + \frac{i\hbar q}{m}\boldsymbol{A}\cdot\nabla + \frac{q^2A^2}{2m} + qV. \tag{13.34}$$

Part of the total potential V will be due to the charges in the atom or molecule, whereas we shall be more concerned with how the Hamiltonian changes when the atom or molecule is in an applied field. We therefore write $\hat{H}$ as the sum of the free molecular Hamiltonian and a field-dependent part $\hat{H}_\text{f}$, so that

$$\hat{H}_\text{f} = \frac{i\hbar q}{m}\boldsymbol{A}\cdot\nabla + \frac{q^2A^2}{2m} + qV_\text{e}, \tag{13.35}$$

where now V_e refers only to the external potential.

13.4 Special cases

Consider first a uniform electric field with no magnetic induction. An electric field of magnitude E along the z direction corresponds to a scalar potential $V = -Ez$, where the zero of potential is taken at the origin of coordinates. The vector potential can be taken as zero, and then

$$\hat{H}_\text{f} = -qEz. \tag{13.36}$$

From eq. (8.6) we see that qz corresponds to the z component of the contribution of the charge q to the electric dipole moment vector $\boldsymbol{p}_\text{e}$, where the subscript e is used to avoid confusion with the momentum. We can therefore write $\hat{H}_\text{f}$ as $-\boldsymbol{p}_\text{e}\cdot\boldsymbol{E}$. This corresponds to the linear term in eq. (8.20) for the energy of a charge distribution in an electrostatic potential with no field gradients and zero potential at the origin.

Conversely, consider a uniform magnetic induction with no electric field.

The scalar potential can be taken as zero. An induction B along the z direction corresponds to

$$A = -\tfrac{1}{2}B(iy - jx). \tag{13.37}$$

There are now two terms in $\hat{H}_f$:

$$\hat{H}_f = \frac{i\hbar qB}{2m}\left(-y\frac{\partial}{\partial x} + x\frac{\partial}{\partial y}\right) + \frac{q^2B^2}{8m}(x^2 + y^2). \tag{13.38}$$

The second term is straightforward in form, although it is quadratic in B rather than linear. The first term can be put in a more compact form by noting that $-i\hbar\partial/\partial x$ is p_x, the x component of momentum. The first term then depends on $xp_y - yp_x$, which is the z component of $\boldsymbol{r} \times \boldsymbol{p}$, and this in turn is the z component $\hat{L}_z$ of the angular momentum operator $\hat{\boldsymbol{L}}$. We can therefore write

$$\hat{H}_f = -\frac{q\hat{L}_z}{2m}B + \frac{q^2B^2}{8m}(\mathbf{x}^2 + y^2) \tag{13.39}$$

From eq. (10.2) we can recognize $q\hat{L}_z/2m$ as the operator $\hat{M}_z$ corresponding to the z component of the magnetic moment, so that another form for H_f is

$$\hat{H}_f = -\hat{M}_zB + (q^2B^2/8m)(\hat{X}^2 + \hat{Y}^2). \tag{13.40}$$

The first term is then the operator equivalent of the energy change of a molecule with a magnetic moment in a magnetic induction, as given in eq. (10.6).

Further reading

P. Landshoff and A. Metherell, *Simple Quantum Physics*, Cambridge University Press, 1979.

J. H. Van Vleck, *The Theory of Electric and Magnetic Susceptibilities*, Oxford University Press, 1965.

D. W. Davies, *The Theory of the Electric and Magnetic Properties of Molecules*, Wiley, London, 1967.

W. P. Healy, *Non-Relativistic Quantum Electrodynamics*, Academic Press, London, 1982.

Chapter 14

Change of quantum state

The electromagnetic properties of molecular systems are of two types. Some describe the electromagnetic state, for example electric and magnetic moments. Others describe the response of the state to electromagnetic perturbations, for example polarizability, magnetizability, and the Einstein coefficients. The present chapter deals with the quantum-mechanical techniques used for calculating changes of quantum state under external perturbations. The perturbation is mostly assumed to be small in some suitable way. The new states and their properties are then described in terms of the old states and their properties in a power series in the strength of the perturbation; for small perturbations the series can be truncated after a few terms. These are the broad features of *perturbation theory*. Both time-independent and time-dependent variants of perturbation theory are met, with differences of emphasis and application.

14.1 Time-independent perturbation theory

The problem to be solved is this: given an unperturbed time-independent Hamiltonian $\hat{H}_0$ for which the (normalized) eigenstates $|n\rangle$ and the eigenvalues E_n are known, what are the eigenstates and eigenvalues of the perturbed Hamiltonian $\hat{H} = \hat{H}_0 + \hat{V}$, where $\hat{V}$ is the perturbation? In fact, the problem of solving one equation in terms of a simpler known one occurs in various areas such as differential equations, and is not unique to quantum mechanics. The unperturbed Schrödinger equation is

$$\hat{H}_0 |n\rangle = E_n |n\rangle. \tag{14.1}$$

Since $\hat{V}$ is supposed small, we assume that the eigenstates of the perturbed Hamiltonian $\hat{H}$ are close to those of $\hat{H}_0$ and can be labelled accordingly as $|\nu_n\rangle$, with energies W_n. (Degenerate eigenstates can change more drastically and thus require a separate treatment given in the next section.) Then the perturbed Schrödinger equation is

$$(\hat{H}_0 + \hat{V}) |\nu_n\rangle = W_n |\nu_n\rangle. \tag{14.2}$$

The new eigenstate and energy are now expanded in a power series:

$$|\nu_n\rangle = |n\rangle + |\nu_n^{(1)}\rangle + |\nu_n^{(2)}\rangle + \dots \tag{14.3}$$

$$W_n = E_n + W_n^{(1)} + W_n^{(2)} + \dots. \tag{14.4}$$

Here $|\nu_n^{(1)}\rangle$ is the first-order correction to $|n\rangle$, proportional to $\hat{V}$ and so increasing by a factor λ if $\hat{V} \to \lambda\hat{V}$; $|\nu_n^{(2)}\rangle$ is the second-order correction, proportional to $\hat{V}^2$ and increasing by λ^2 if $\hat{V} \to \lambda\hat{V}$; and so on. Similarly, $W_n^{(1)}$ is the first-order correction to E_n, and so on. The unperturbed Hamiltonian, eigenstates and energies are of course independent of $\hat{V}$, thus increasing by a factor $\lambda^0 = 1$ if $\hat{V} \to \lambda\hat{V}$.

The expressions (14.3) and (14.4) are now substituted in eq. (14.2) to give

$$(\hat{H}_0 + \hat{V})[|n\rangle + |\nu_n^{(1)}\rangle + |\nu_n^{(2)}\rangle + \dots]$$
$$= [E_n + W_n^{(1)} + W_n^{(2)} + \dots][|n\rangle + |\nu_n^{(1)}\rangle + |\nu_n^{(2)}\rangle \dots]. \tag{14.5}$$

The two sides of this equation are power series in the magnitude of $\hat{V}$. The only way the two sides can be equal (except possibly for certain particular magnitudes of $\hat{V}$) is for the coefficients of corresponding order on each side to be equal. Terms of different order change in different proportions with $\hat{V}$ and so cannot be equal. In zeroth order, i.e. for terms independent of $\hat{V}$, we find

$$\hat{H}_0|n\rangle = E_n|n\rangle. \tag{14.6}$$

This is the unperturbed equation, which is clearly all that can be independent of $\hat{V}$ in general. In first order we find

$$\hat{H}_0|\nu_n^{(1)}\rangle + \hat{V}|n\rangle = E_n|\nu_n^{(1)}\rangle + W_n^{(1)}|n\rangle \tag{14.7}$$

and in second order

$$\hat{H}_0|\nu_n^{(2)}\rangle + \hat{V}|\nu_n^{(1)}\rangle = E_n|\nu_n^{(2)}\rangle + W_n^{(1)}|\nu_n^{(1)}\rangle + W_n^{(2)}|n\rangle. \tag{14.8}$$

Only in special cases is it necessary to go beyond second order, and first order often suffices.

The first-order correction to the energy is $W_n^{(1)}$. It is obtained from eq. (14.7) by forming the scalar product of both sides with $\langle n|$, using the fact that $\langle n|n\rangle = 1$ for normalized states, so that

$$\langle n|\hat{H}_0|\nu_n^{(1)}\rangle + \langle n|\hat{V}|n\rangle = E_n\langle n|\nu_n^{(1)}\rangle + W_n^{(1)}. \tag{14.9}$$

Because $\hat{H}_0$ is a Hermitian operator, the first term on the left-hand side can be rewritten as follows:

$$\langle n|\hat{H}_0|\nu_n^{(1)}\rangle = \langle \nu_n^{(1)}|\hat{H}_0|n\rangle^* \tag{14.10}$$

$$= (E_n\langle \nu_n^{(1)}|n\rangle)^* \tag{14.11}$$

$$= E_n\langle n|\nu_n^{(1)}\rangle, \tag{14.12}$$

where the fact that E_n is real has been used. We therefore obtain the simple result

$$W_n^{(1)} = \langle n \,|\, \hat{V} \,|\, n \rangle. \tag{14.13}$$

Comparison with the unperturbed result

$$E_n = \langle n \,|\, \hat{H}_0 \,|\, n \rangle \tag{14.14}$$

shows that to first order the new energy is given by the expectation value of the new Hamiltonian in the old unperturbed state. To first order, the energy change does not depend on any distortion of the state.

The first-order correction to the state $|\, n \rangle$ is expressed in terms of an admixture of contributions from other states $|\, m \rangle$. As these form a complete set, the correction can be expanded as

$$|\, \nu_n^{(1)} \rangle = \sum_m c_{nm} |\, m \rangle, \tag{14.15}$$

where $c_{nm} = \langle m \,|\, \nu_n^{(1)} \rangle$. These coefficients can be deduced from eq. (14.7) by forming the scalar product of both sides with $\langle m \,|$, using the fact that $\langle m \,|\, n \rangle = 0$ for $m \neq n$ (i.e. the states are orthogonal, which can always be ensured even if they are degenerate). We find

$$\langle m \,|\, \hat{H}_0 \,|\, \nu_n^{(1)} \rangle + \langle m \,|\, \hat{V} \,|\, n \rangle = E_n \langle m \,|\, \nu_n^{(1)} \rangle, \tag{14.16}$$

where use of eq. (14.12) allows us to write

$$(E_n - E_m) c_{nm} = \langle m \,|\, \hat{V} \,|\, n \rangle \equiv V_{mn}. \tag{14.17}$$

Provided the states are non-degenerate so that $E_n \neq E_m$, we can obtain c_{nm} for $n \neq m$ as

$$c_{nm} = V_{mn} / (E_n - E_m). \tag{14.18}$$

It is clear that this cannot be correct if the states are degenerate, since then the coefficients would scarcely be small as expected, which is why the different treatment of the next section is required.

The coefficient c_{nn} is obtained from the requirement that the new state $|\, \nu_n \rangle$ should be normalized, i.e.

$$\langle \nu_n \,|\, \nu_n \rangle = 1. \tag{14.19}$$

Substitution from eq. (14.3) then yields

$$\begin{aligned} &\langle n \,|\, n \rangle + (\langle n \,|\, \nu_n^{(1)} \rangle + \langle \nu_n^{(1)} \,|\, n \rangle) \\ &\quad + (\langle n \,|\, \nu_n^{(2)} \rangle + \langle \nu_n^{(1)} \,|\, \nu_n^{(1)} \rangle + \langle \nu_n^{(2)} \,|\, n \rangle) \\ &\quad + \ldots = 1, \end{aligned} \tag{14.20}$$

where terms of the same order have been bracketed together. Since $\langle n \,|\, n \rangle = 1$ already, this equation requires that each bracket should be zero. In particular, the first-order terms imply

$$\mathrm{Re}\,\langle n \mid v_n^{(1)} \rangle = 0 \tag{14.21}$$

since $\langle v_n^{(1)} \mid n \rangle = \langle n \mid v_n^{(1)} \rangle^*$. Using eq. (14.15) for $\mid v_n^{(1)} \rangle$ we find that since $\langle n \mid m \rangle = 0$ unless $n = m$, then $\mathrm{Re}\, c_{nn} = 0$. If $\mid n \rangle$ and $\mid v_n^{(1)} \rangle$ are also real, then $\mathrm{Im}\, c_{nn} = 0$ too, and we have $c_{nn} = 0$. Thus the first-order correction to the state $\mid n \rangle$ consists of mixing in *other* states according to

$$\mid v_n^{(1)} \rangle = \sum_{m(\neq n)} \frac{V_{mn}}{E_n - E_m} \mid m \rangle. \tag{14.22}$$

For a given perturbation matrix element V_{mn}, the further away in energy state $\mid m \rangle$ is from state $\mid n \rangle$, the less it contributes to the perturbation of state $\mid n \rangle$.

The second-order correction to the energy is sometimes required, for example when the first-order correction vanishes identically by symmetry. Forming the scalar product of both sides of eq. (14.8) with $\langle n \mid$, we obtain

$$\langle n \mid \hat{H}_0 \mid v_n^{(2)} \rangle + \langle n \mid \hat{V} \mid v_n^{(1)} \rangle$$
$$= E_n \langle n \mid v_n^{(2)} \rangle + W_n^{(1)} \langle n \mid v_n^{(1)} \rangle + W_n^{(2)}. \tag{14.23}$$

Here the first terms on each side are equal, by eq. (14.12), and the coefficient of $W_n^{(1)}$ is c_{nn}, which is zero. This leaves

$$W_n^{(2)} = \langle n \mid \hat{V} \mid v_n^{(1)} \rangle, \tag{14.24}$$

which therefore now does involve distortion of the initial state. Substitution for $\mid v_n^{(1)} \rangle$ from eq. (14.22) yields

$$W_n^{(2)} = \sum_{m(\neq n)} \frac{V_{nm} V_{mn}}{E_n - E_m} \tag{14.25}$$

$$= \sum_{m(\neq n)} \frac{\mid V_{nm} \mid^2}{E_n - E_m}. \tag{14.26}$$

Only in a few special cases can the sum in these equations be carried out exactly (in which cases the exact result can usually be arrived at more directly than by perturbation theory). Therefore, it is useful to be able to estimate $W_n^{(2)}$. This is often done by setting $E_n - E_m$ equal to some average excitation energy Δ. Then

$$W_n^{(2)} \approx \left(\sum_m V_{nm} V_{mn} - V_{nn}^2 \right) \Big/ \Delta, \tag{14.27}$$

where the restriction on the sum over m has been dropped and the term for $m = n$ subtracted out. The sum is then the nn matrix element of $\hat{V}^2$, so that

$$W_n^{(2)} \approx [(\hat{V}^2)_{nn} - V_{nn}^2]/\Delta. \tag{14.28}$$

In this approximation, $W_n^{(2)}$ is proportional to the variance of $\hat{V}$ about its mean value in the state n:

$$W_n^{(2)} \approx \langle n \mid (\hat{V} - V_{nn})^2 \mid n \rangle/\Delta. \tag{14.29}$$

This result shows that if the variance is small compared with the product ΔV_{nn}, then $W_n^{(2)} \ll W_n^{(1)}$ and the perturbation series for the energy is rapidly converging.

14.2 Degenerate perturbation theory

When two states $|n\rangle$ and $|m\rangle$ are degenerate, the foregoing treatment for $|v_n^{(1)}\rangle$ and $W_n^{(2)}$ breaks down, because the coefficient c_{nm} diverges (unless V_{nm} also vanishes, which is not generally so). This means that a small admixture of the state $|m\rangle$ is inadequate to represent $|v_n^{(1)}\rangle$, so that the very idea of a perturbation series is inappropriate, at least in the previous form. The origin of this problem, and at the same time the means of its solution, is the fact that any linear combination of degenerate states $|n\rangle$, $|n'\rangle$, etc., all of energy E_n, is also an eigenstate of energy E_n. We must therefore allow the perturbation to form different linear combinations as well as distorting the states.

Suppose for simplicity that just the two states $|n\rangle$ and $|n'\rangle$ are degenerate, but are chosen to be orthogonal. The perturbed state $|v_n\rangle$ now has to be written as

$$|v_n\rangle = a_{nn}|n\rangle + a_{nn'}|n'\rangle + |v_n^{(1)}\rangle + |v_n^{(2)}\rangle + \ldots \qquad (14.30)$$

The first-order equation (14.7) then becomes

$$\hat{H}_0|v_n^{(1)}\rangle + \hat{V}(a_{nn}|n\rangle + a_{nn'}|n'\rangle) = E_n|v_n^{(1)}\rangle + W_n^{(1)}(a_{nn}|n\rangle + a_{nn'}|n'\rangle). \qquad (14.31)$$

We now form the scalar product of this equation with $\langle n|$, as in obtaining eq. (14.9). From eq. (14.12) the first terms on each side of the equation are then equal. Since $\langle n|n'\rangle = 0$, on rearrangement we find

$$a_{nn}[V_{nn} - W_n^{(1)}] + a_{nn'}V_{nn'} = 0. \qquad (14.32)$$

Similarly, forming the scalar product of both sides of eq. (14.31) with $\langle n'|$ leads to

$$a_{nn}V_{n'n} + a_{nn'}[V_{n'n'} - W_n^{(1)}] = 0. \qquad (14.33)$$

The coefficients a_{nn} and $a_{nn'}$ describing the mixing of the zeroth-order states are yet to be determined. They are the solutions of the linear homogeneous equations (14.32) and (14.33). However, such equations have non-zero solutions only if the determinant of the coefficients vanishes, i.e. in this case if

$$\begin{vmatrix} V_{nn} - W_n^{(1)} & V_{nn'} \\ V_{n'n} & V_{n'n'} - W_n^{(1)} \end{vmatrix} = 0. \qquad (14.34)$$

Expansion of the determinant yields a quadratic equation for the first order change in the energy:

$$[W_n^{(1)}]^2 - (V_{nn} + V_{n'n'})W_n^{(1)} + (V_{nn}V_{n'n'} - V_{nn'}V_{n'n}) = 0. \qquad (14.35)$$

This has two solutions, as one expects, namely

$$W_n^{(1)} = \tfrac{1}{2}\{V_{nn} + V_{n'n'} \pm [(V_{nn} - V_{n'n'})^2 + 4\,|\,V_{nn'}\,|^2]^{\frac{1}{2}}\}, \qquad (14.36)$$

which actually correspond to the energy changes for the two degenerate states, as can be verified by carrying the foregoing procedure through for state $|\,v_{n'}\rangle$ instead of $|\,v_n\rangle$. In general the degeneracy of the states is lifted, the exception occurring when $V_{nn} = V_{n'n'}$ and $|\,V_{nn'}\,| = 0$. It can also be seen that the term in $|\,V_{nn'}\,|^2$ causes the first-order energy change to differ from the non-degenerate results V_{nn} and $V_{n'n'}$ for the two states. This is illustrated in Fig. 14.1, showing how the interaction $V_{nn'}$ through the perturbation causes the energy levels to 'repel'.

For either of the perturbation energies $W_n^{(1)}$, eqs (14.32) and (14.33) give the ratio $a_{nn}/a_{nn'}$. The individual coefficients are then determined from the normalization $\langle v_n\,|\,v_n\rangle = 1$. From eq. (14.30) the scalar product is expanded and the zeroth-order terms equated to unity. Since $\langle n\,|\,n'\,\rangle = 0$ already, we obtain

$$|\,a_{nn}\,|^2 + |\,a_{nn'}\,|^2 = 1, \qquad (14.37)$$

which fixes the coefficients.

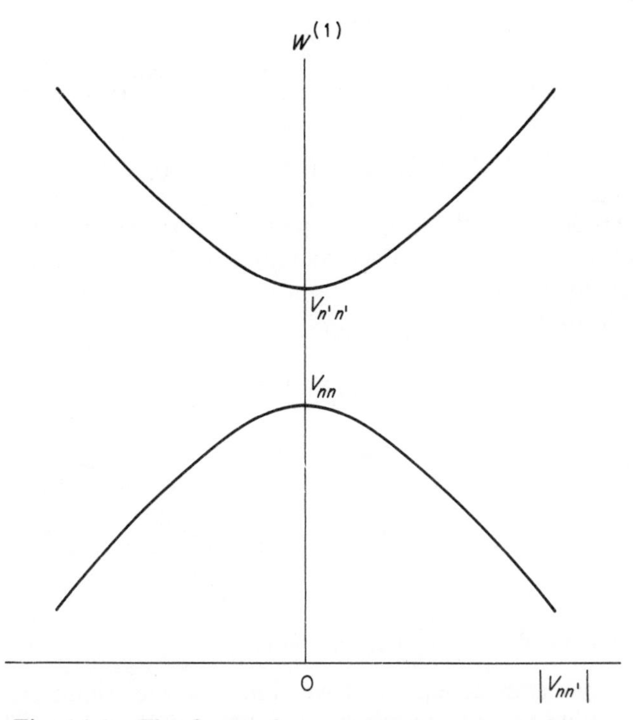

Fig. 14.1 The first-order perturbation energy for a pair of degenerate states $|\,n\rangle$ and $|\,n'\,\rangle$ as a function of the interaction matrix element $V_{nn'}$

Once the appropriate linear combination of the degenerate zeroth-order states is formed, the first-order correction proceeds much as before. We expand

$$|\,v_n^{(1)}\rangle = \sum_{m(\neq n,n')} c_{nm}\,|\,m\rangle, \tag{14.38}$$

substitute in eq. (14.31) and form the scalar product of both sides with $\langle m\,|$. This yields

$$(E_n - E_m)c_{nm} = a_{nn}V_{mn} + a_{nn'}V_{mn'}, \tag{14.39}$$

and since $E_m \neq E_n$ by definition, we obtain

$$c_{nm} = (a_{nn}V_{mn} + a_{nn'}V_{mn'})/(E_n - E_m). \tag{14.40}$$

Other results for the states $|\,n\rangle$ and $|\,n'\rangle$ follow with similar modifications. Perturbation theory for the other states $|\,m\rangle$ follows exactly as in the non-degenerate case.

Degeneracies do not always occur in pairs, of course. For a set of q degenerate states labelled r, the foregoing undergoes straightforward modifications. The zeroth-order part of $|\,v_n\rangle$ is expanded as

$$|\,v_r^{(0)}\rangle = \sum_{r'} a_{rr'}\,|\,r'\rangle, \tag{14.41}$$

leading to q equations for the coefficients $a_{rr'}$. Non-zero solutions require the vanishing of a qth order determinant

$$\det\|\,V_{rr'} - W_r^{(1)}\,\delta_{rr'}\,\| = 0. \tag{14.42}$$

This has q roots giving the first-order perturbation energies, which may partly or wholly lift the degeneracy. Each root gives $q-1$ independent equations for the coefficients $a_{rr'}$, which are determined with the additional equation from the normalization

$$\sum_{r'} |\,a_{rr'}\,|^2 = 1. \tag{14.43}$$

The first-order change in the wavefunction is given in terms of states $|\,m\rangle$ outside the set $\{|\,r\rangle\}$ by coefficients c_{rm} which are found from

$$c_{rm} = \sum_{r'} a_{rr'}V_{mr'}/(E_r - E_m). \tag{14.44}$$

14.3 Time-dependent perturbation theory

In the previous two sections we have seen how a constant perturbation $\hat{V}$ changes the state and energy of a system. We now investigate what happens when the perturbation is time-dependent. Strictly speaking, all perturbations that we can apply must be time-dependent: we cannot arrange for them to have been in existence from a time $t = -\infty$, and must instead switch them on. As in an electrical circuit, such switching on causes initial transient behaviour which eventually dies away to leave a steady rate of flow in suitable cases. The results of time-independent perturbation theory can, however, be regained in a treatment in which the perturbation is switched on sufficiently slowly or *adiabatically*. This mixes the unperturbed states together to give the perturbed states without causing the transitions between unperturbed states which are the principal concern of time-dependent perturbation theory.

We now suppose that the perturbed system is described by the Hamiltonian

$$\hat{H}(t) = \hat{H}_0 + \hat{V}(t), \tag{14.45}$$

where the eigenstates and eigenvalues of $\hat{H}_0$ are as in eq. (14.1):

$$\hat{H}_0 \,|\, n\rangle = E_n \,|\, n\rangle. \tag{14.46}$$

The corresponding time-dependent states are of the form

$$|\, n,\, t\rangle = |\, n\rangle\, \mathrm{e}^{-iE_n t/\hbar}, \tag{14.47}$$

as in Chapter 11, where it can be seen that $|\, n\rangle = |\, n,\, 0\rangle$. The state of the perturbed system is written as $|\, v(t)\rangle$, which satisfies the time-dependent Schrödinger equation (11.17)

$$i\hbar\, \partial \,|\, v(t)\rangle/\partial t = \hat{H}(t) \,|\, v(t)\rangle. \tag{14.48}$$

In the same spirit as before, $|\, v(t)\rangle$ is expanded as a series of contributions from the time-dependent unperturbed states, so that

$$|\, v(t)\rangle = \sum_n c_n(t) \,|\, n,\, t\rangle. \tag{14.49}$$

Here $c_n(t) = \langle n,\, t \,|\, v(t)\rangle$ and $|\, c_n(t)\,|^2$ can be regarded as the fraction of the state $|\, n,\, t\rangle$ contained by the perturbed state $|\, v(t)\rangle$, which must, of course, be allowed to depend on time, in general.

The expansion (14.49) is substituted in eq. (14.48) with eq. (14.45) for the Hamiltonian to yield

$$i\hbar \frac{\partial}{\partial t} \sum_n c_n(t) \,|\, n,\, t\rangle = [\hat{H}_0 + \hat{V}(t)] \sum_n c_n(t) \,|\, n,\, t\rangle \tag{14.50}$$

whence

$$i\hbar \sum_n \left[\frac{\partial c_n(t)}{\partial t} \,|\, n,\, t\rangle + c_n(t) \frac{\partial}{\partial t} \,|\, n,\, t\rangle \right] = \sum_n c_n(t) [\hat{H}_0 + \hat{V}(t)] \,|\, n,\, t\rangle. \tag{14.51}$$

This equation contains the time-dependent Schrödinger equation for the unperturbed states

$$i\hbar \frac{\partial}{\partial t} \,|\, n,\, t\rangle = \hat{H}_0 \,|\, n,\, t\rangle, \tag{14.52}$$

leaving

$$i\hbar \sum_n \frac{\partial c_n(t)}{\partial t} | n, t\rangle = \sum_n c_n(t) \hat{V}(t) | n, t\rangle. \tag{14.53}$$

The time derivative of the coefficient $c_m(t)$ for a particular state $|m, t\rangle$ is obtained by forming the scalar product of both sides of this equation with

$$\langle m, t| = \langle m | e^{iE_m t/h} \tag{14.54}$$

and using the orthogonality of the states $|m\rangle$ and $|n\rangle$. Thus

$$i\hbar \frac{\partial c_m(t)}{\partial t} = \sum_n c_n(t) \langle m, t | \hat{V}(t) | n, t\rangle, \tag{14.55}$$

where the matrix element of the perturbation on the right-hand side can be written as

$$\langle m, t | \hat{V}(t) | n, t\rangle = \langle m | \hat{V}(t) | n\rangle e^{i(E_m - E_n)t/h}. \tag{14.56}$$

Writing $\langle m | \hat{V}(t) | n\rangle = V_{mn}(t)$ and $E_m - E_n = \hbar\omega_{mn}$, we obtain the result

$$\partial c_m(t)/\partial t = (1/i\hbar) \sum_n c_n(t) V_{mn}(t) e^{i\omega_{mn} t}. \tag{14.57}$$

This is exact: so far no approximations have been introduced, and we can obtain an expression for $c_m(t)$ by integrating:

$$c_m(t_2) = c_m(t_1) + (1/i\hbar) \sum_n \int_{t_1}^{t_2} c_n(t) V_{mn}(t)\, e^{i\omega_{mn} t}\, dt. \tag{14.58}$$

The problem is that this integral equation is seldom easy to solve. A general solution can, however, be obtained to first order in the magnitude of the perturbation.

Suppose that the system was in the state $|k\rangle$ at time zero, i.e. $|\nu(0)\rangle = |k, 0\rangle$, and expand $c_n(t)$ in a power series in the magnitude of $\hat{V}(t)$:

$$c_n(t) = \delta_{nk} + c_n^{(1)}(t) + c_n^{(2)}(t) + \ldots. \tag{14.59}$$

Here $c_n^{(1)}(t)$ varies like $\hat{V}(t)$, $c_n^{(2)}(t)$ like $\hat{V}(t)^2$, and so on, and all these correction terms are zero at $t = 0$. From eq. (14.59) it also follows that

$$\partial c_n(t)/\partial t = \partial c_n^{(1)}(t)/\partial t + \partial c_n^{(2)}(t)/\partial t + \ldots. \tag{14.60}$$

This is substituted on the left-hand side of eq. (14.57), and the expansion (14.59) is substituted on the right-hand side. There are no zeroth-order terms, but the first-order terms yield

$$\partial c_m^{(1)}/\partial t = (1/i\hbar) \sum_n \delta_{nk} V_{mn}(t) e^{i\omega_{mn} t}, \tag{14.61}$$

where the first-order factor $V_{mn}(t)$ has to be combined with the zeroth-order term in $c_n(t)$, namely δ_{nk}. In the sum over n in this equation, only the term $n = k$ survives, leaving

$$\partial c_m^{(1)}/\partial t = (1/i\hbar) V_{mk}(t) e^{i\omega_{mk} t}, \tag{14.62}$$

which can be integrated to yield the explicit expression

$$c_m^{(1)}(t) = (1/i\hbar)\int_0^t V_{mk}(t')\mathrm{e}^{i\omega_{mk}t'}\,\mathrm{d}t'. \tag{14.63}$$

Similar expressions can be derived for the higher-order terms $c_m^{(r)}(t)$. However, so long as the probability of occupation of the initial state $|k\rangle$ remains close to unity, the probability amplitude $c_m(t)$ can be approximated by its first-order term $c_m^{(1)}(t)$ (for $m \neq k$). This implies that none of the other states $|m'\rangle$ is sufficiently populated to allow the state $|m\rangle$ to be populated by a second-order process from $|k\rangle$ via $|m'\rangle$ instead of by a first-order process directly from $|k\rangle$. Henceforth we shall adopt this approximation, dropping the superscript on $c_m^{(1)}(t)$ and taking $|c_m^{(1)}(t)|^2$ as the probability that the system has made a transition from the state $|k\rangle$ to the state $|m\rangle$ after time t. This can be obtained explicitly from eq. (14.63) for particular forms of perturbation $\hat{V}(t)$.

14.4 Constant perturbation

We now consider the effect of the special perturbation mentioned at the begining of the last section, which consists of switching on an otherwise constant perturbation. The corresponding perturbation operator is

$$\begin{aligned} \hat{V}(t') &= \hat{V} \qquad (0 \leqslant t' \leqslant t) \\ &= 0 \qquad (\text{otherwise}), \end{aligned} \tag{14.64}$$

where t is the time of observation. The probability amplitude $c_m(t)$ follows from eq. (14.63) as

$$c_m(t) = (V_{mk}/i\hbar)\int_0^t \mathrm{e}^{i\omega_{mk}t'}\,\mathrm{d}t' \tag{14.65}$$

$$= (V_{mk}/\hbar\omega_{mk})(1 - \mathrm{e}^{i\omega_{mk}t}). \tag{14.66}$$

The probability of finding the system in state $|m\rangle$ at time t is

$$P(k \to m) = |c_m(t)|^2 \tag{14.67}$$

$$= (|V_{mk}|/\hbar\omega_{mk})^2(2 - \mathrm{e}^{i\omega_{mk}t} - \mathrm{e}^{-i\omega_{mk}t}) \tag{14.68}$$

$$= 2(|V_{mk}|/\hbar\omega_{mk})^2(1 - \cos\omega_{mk}t) \tag{14.69}$$

which it is convenient to transform by the identity $\cos 2\theta = 1 - 2\sin^2\theta$ into

$$P(k \to m) = \frac{|V_{mk}|^2}{\hbar^2}\left\{\frac{\sin^2\frac{1}{2}\omega_{mk}t}{(\frac{1}{2}\omega_{mk})^2}\right\}. \tag{14.70}$$

The behaviour of the trigonometric function in curly brackets is sketched in Fig. 14.2. It has zeros where $\frac{1}{2}\omega_{mk}t = n\pi$ for $n = \pm 1, \pm 2, \ldots$ However, as $t \to 0$ the function tends to t^2. The area under the curve is given by a standard integral as $2\pi t$, which also equals the product of the height t^2 of the central

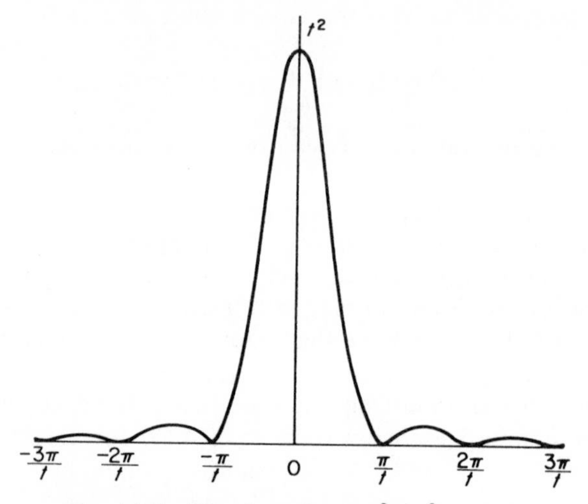

Fig. 14.2 The function $x^{-2}\sin^2 xt$ as a function of x. For t large compared with $1/x$, the central peak becomes very steep and the subsidiary maxima approach the origin more closely

peak and the width $2\pi/t$ at half height. For times $t \gg 1/\omega_{mk}$, the central peak becomes very high and narrow, the function thus being different from zero only in a very small neighbourhood around $\omega_{mk} = 0$. In this limit the function tends to

$$2\pi t\ \delta(\omega_{mk}) = 2\pi\hbar t\ \delta(E_m - E_k), \tag{14.71}$$

where $\delta(x)$ is the Dirac delta function. Because this function is infinitely sharp at $x = 0$ and zero elsewhere, it has the effect under an integral sign of picking out the value of the rest of the integrand at $x = 0$, all other points contributing nothing. In particular, the integral of $\delta(x)$ itself is unity over any range including the origin.

The transition probability then becomes

$$P(k \to m) = (2\pi t \mid V_{mk} \mid^2/\hbar)\ \delta(E_m - E_k). \tag{14.72}$$

Since this probability is proportional to the time t, we obtain a constant *transition rate*

$$W(k \to m) = \mathrm{d}P(k \to m)/\mathrm{d}t \tag{14.73}$$

$$= (2\pi \mid V_{mk} \mid^2/\hbar)\ \delta(E_m - E_k). \tag{14.74}$$

The delta function is non-zero only where $E_m = E_k$, and hence shows that transitions occur only to states $\mid m\rangle$ of the same energy as the initial state $\mid k\rangle$, at least within a range of the order of $\hbar/t$. The delta function can therefore be taken as expressing energy conservation.

It often happens that the final state $|m\rangle$ is one of a set closely spaced in energy, all with similar values for $|V_{mk}\rangle$. For example, the set might comprise vibrational and rotational sublevels of an electronic level. Let the density of these states per unit energy be $\varrho(E_m)$, that is, there are $\varrho(E_m)\,\mathrm{d}E_m$ states in the interval between E_m and $E_m + \mathrm{d}E_m$. Then the total rate of transition out of the initial state $|k\rangle$ to the set of final states is

$$W = \int W(k \rightarrow m)\,\varrho(E_m)\,\mathrm{d}E_m \tag{14.75}$$

$$= (2\pi/\hbar)\int |V_{mk}|^2\,\varrho(E_m)\,\delta(E_m - E_k)\,\mathrm{d}E_m. \tag{14.76}$$

If $|V_{mk}|$ is independent of E_m, it can be taken outside the integral, which because of the delta function is equal to the integrand with $E_m = E_k$, leaving

$$W = (2\pi/\hbar)\,|V_{mk}|^2\,\varrho(E_k). \tag{14.77}$$

This important result is known as *Fermi's Golden Rule*. It is widely used in calculating rates of processes such as the absorption and emission of photons, radiationless transitions, etc. (see Chapter 16).

The importance of the quasi-continuum of closely-spaced final states is that it ensures irreversibility. The conditions that favour the transition $k \rightarrow m$ equally favour the reverse transition $m \rightarrow k$, and indeed for a pair of isolated states the probability of occupation of either state oscillates in time (see Section 14.7). Note, however, that the final states must be described by an extra term $\hat{V}'$ in the total Hamiltonian such that $\hat{H}_0 \gg \hat{V}' \gg \hat{V}$, and then processes dependent on $\hat{V}$ yield the rate-determining step.

14.5 Periodic perturbation

Another important case of time-dependent perturbation theory arises with the periodic perturbation, switched on at $t = 0$,

$$\hat{V}(t) = 2\hat{V}\cos\omega t \tag{14.78}$$

$$= \hat{V}(\mathrm{e}^{i\omega t} + \mathrm{e}^{-i\omega t}). \tag{14.79}$$

Other time-dependent perturbations can be expressed as Fourier series of terms like this. From eq. (14.63) the probability amplitude is

$$c_m(t) = (V_{mk}/i\hbar)\int_0^t [\mathrm{e}^{i(\omega_{mk}+\omega)t'} + \mathrm{e}^{i(\omega_{mk}-\omega)t'}]\,\mathrm{d}t' \tag{14.80}$$

$$= \frac{V_{mk}}{i\hbar}\left[\frac{\mathrm{e}^{i(\omega_{mk}+\omega)t} - 1}{i(\omega_{mk}+\omega)} + \frac{\mathrm{e}^{i(\omega_{mk}-\omega)t} - 1}{i(\omega_{mk}-\omega)}\right]. \tag{14.81}$$

Each term here is like one for the constant perturbation but with ω_{mk} replaced by $\omega_{mk} \pm \omega$. Moreover, the magnitude of $c_m(t)$ is clearly largest when ω_{mk} is close to $\pm\omega$, in which case one term far outweighs the other. Then in calculating the transition probability $P(k \rightarrow m)$ the cross terms between the two parts of eq. (14.81) can be neglected, leaving two terms like eq. (14.72):

$$P(k \to m) = (2\pi t\,|\,V_{mk}\,|^2/\hbar)[\delta(E_m - E_k + \hbar\omega) + \delta(E_m - E_k - \hbar\omega)]\,. \quad (14.82)$$

The rate $W(k \to m)$ follows directly, and the total rate of transition to a set of final states of density $\varrho(E_m)$ is given by

$$W = (2\pi/\hbar)\,|\,V_{mk}\,|^2[\varrho(E_k + \hbar\omega) + \varrho(E_k - \hbar\omega)]\,. \quad (14.83)$$

Transitions are possible to states with energies $E_m = E_k \pm \hbar\omega$, that is, with energies differing by one quantum from the initial state, e.g. one photon. This result also shows how a periodic perturbation stimulates both absorption and emission, as with the Einstein coefficients. The rates of absorption and emission depend on the densities of final states at energies $E_k + \hbar\omega$ and $E_k - \hbar\omega$ respectively.

The results in this section are readily modified to show the effect when a perturbation is switched on adiabatically. We take

$$\hat{V}(t) = \hat{V}e^{\gamma t}, \quad (14.84)$$

where γ is a small positive quantity so that $\hat{V}(t)$ is zero at $t = -\infty$ and equal to $\hat{V}$ at $t = 0$. This perturbation is equivalent to the periodic perturbation $e^{-i\omega t}$ but with an imaginary frequency $i\gamma$. Then by analogy with eq. (14.80) we find

$$c_m(t) = (V_{mk}/i\hbar)\int_{-\infty}^{t} e^{i(\omega_{mk} - i\gamma)t'}\,dt'\,. \quad (14.85)$$

At $t = -\infty$ the integrand vanishes, leaving

$$c_m(t) = (V_{mk}/i\hbar)\,\frac{e^{i(\omega_{mk} - i\gamma)t}}{i(\omega_{mk} - i\gamma)}. \quad (14.86)$$

We now make γ small enough so that $\gamma t \ll 1$ and $\gamma \ll \omega_{mk}$, which can always be done for any time of observation, and then

$$c_m(t) = -\,V_{mk}e^{i\omega_{mk}t}/\hbar\omega_{mk}. \quad (14.87)$$

This corresponds to the coefficient c_{km} given by eq. (14.18) in the time-independent case, except that here the expansion includes the time dependence of the zeroth-order states to give the factor $e^{i\omega_{mk}t}$. In particular, $|\,c_m(t)\,|^2$ is equal to the square of the time-independent coefficient $|\,c_{km}\,|^2$.

14.6 Lifetime of excited states

The Einstein coefficients show that the rate of decay of an upper state containing N_u molecules is of the form

$$-\,dN_u/dt = \Gamma N_u. \quad (14.88)$$

This is a simple first-order rate law with the solution

$$P(t) = N_u(t)/N_u(0) = e^{-\Gamma t}. \quad (14.89)$$

Here $P(t)$ is the probability that the upper state is still occupied at time t, and Γ is the inverse *lifetime* of the state ($1/\Gamma$ is the time taken for a fraction $1/e$ of the population to decay). We now use this result to modify the time-dependent perturbation theory to allow for the depletion of the initial state at the rate Γ. Making the results self-consistent will yield an expression for Γ.

Now if the initial state is again $|k, 0\rangle$, we have

$$P(t) = |c_k(t)|^2 = e^{-\Gamma t}. \tag{14.90}$$

Therefore in the perturbation expansion (14.49) for $c_n(t)$ we replace the zeroth-order term δ_{nk} by

$$c_n^{(0)}(t) = \delta_{nk} e^{-\frac{1}{2}\Gamma t}. \tag{14.91}$$

The first-order perturbation treatment follows as before, except that the exponential $e^{i\omega_{mn}t}$ in eq. (14.62) is replaced by

$$e^{i\omega_{mk}t}\, e^{-\frac{1}{2}\Gamma t} = e^{i(\omega_{mk} + \frac{1}{2}i\Gamma)t}. \tag{14.92}$$

Since emission processes are being considered here, $E_m < E_k$, so that $\omega_{mk} < 0$ or $\omega_{km} > 0$. Emission also implies the periodic perturbation

$$\hat{V}(t) = \hat{V} e^{i\omega t}. \tag{14.93}$$

Then using the results of the previous section with ω_{mk} replaced by $-\omega_{km} + \frac{1}{2}i\Gamma$, we find

$$c_m(t) = \frac{V_{mk}}{i\hbar} \frac{e^{i(\omega - \omega_{km} + \frac{1}{2}i\Gamma)t} - 1}{i(\omega - \omega_{km} + \frac{1}{2}i\Gamma)}. \tag{14.94}$$

After a time $t \gg 1/\Gamma$ the factor $e^{-\frac{1}{2}\Gamma t}$ causes the time-dependent term to decay to a negligible magnitude, leaving constant probability amplitudes

$$c_m(\infty) = V_{mk}/\hbar(\omega - \omega_{mk} + \tfrac{1}{2}i\Gamma). \tag{14.95}$$

The probability of transition from the initial state $|k\rangle$ to the lower state $|m\rangle$ after several lifetimes have elapsed is thus

$$|c_m(\infty)|^2 = \frac{|V_{mk}|^2}{(\hbar\omega - E_k + E_m)^2 + \frac{1}{4}\hbar^2\Gamma^2} \tag{14.96}$$

$$= \frac{|V_{mk}|^2}{\hbar\Gamma} \left\{ \frac{\hbar\Gamma}{(\hbar\omega - E_k + E_m)^2 + \frac{1}{4}\hbar^2\Gamma^2} \right\}. \tag{14.97}$$

The function in curly brackets here is sketched in Fig. 14.3; it is known as a *Lorentzian* function. It has a maximum value of $4/\hbar\Gamma$ at $E_m = E_k - \hbar\omega$, and reaches its half maximum value at $E_m = E_k - \hbar\omega \pm \frac{1}{2}\hbar\Gamma$, so that its full width at half maximum is $\hbar\Gamma$. Its integral with respect to $\hbar\omega$ is 2π. As $\Gamma \to 0$ the integral is unchanged, but the function becomes increasingly high, sharp and narrow. It therefore tends to $2\pi\, \delta(E_k - E_m - \hbar\omega)$ as Γ tends to zero.

If $\varrho(E_m)$ is the density of states at the final energy, the total probability of

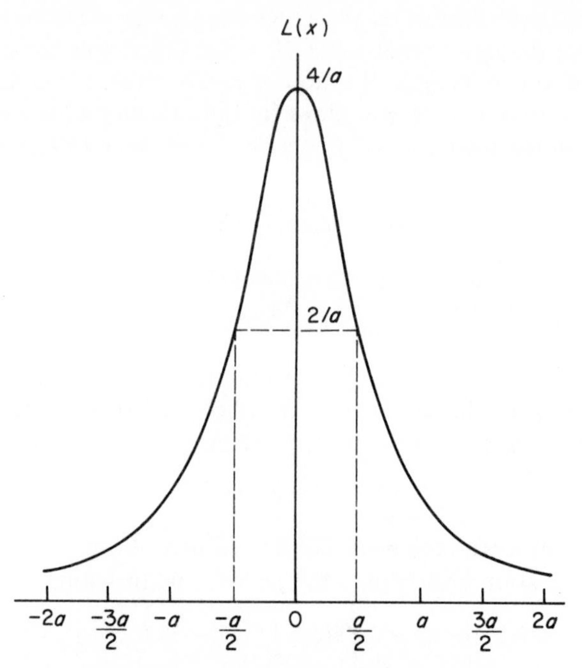

Fig. 14.3 The Lorentzian function $L(x) = a/(x^2 + \frac{1}{4}a^2)$. The full width at half-height is a as indicated

transition is

$$\int |c_m(\infty)|^2 \varrho(E_m)\, \mathrm{d}E_m \approx \frac{2\pi |V_{mk}|^2}{\hbar\Gamma} \varrho(E_k - \hbar\omega) \tag{14.98}$$

for sufficiently small Γ (long lifetime). If we sum over all states $|m\rangle$ with $E_m < E_k$, the probabilities must sum to unity: at long enough times, the probability $\mathrm{e}^{-\Gamma t}$ of remaining in the initial state tends to zero and the probability of being in some lower state therefore tends to unity. We therefore have

$$1 = (2\pi/\hbar\Gamma) \sum_{m(E_m < E_k)} |V_{mk}|^2 \varrho(E_k - \hbar\omega), \tag{14.99}$$

from which the decay rate is obtained as

$$\Gamma = (2\pi/\hbar) \sum_{m(E_m < E_k)} |V_{mk}|^2 \varrho(E_k - \hbar\omega). \tag{14.100}$$

Compare this result with Fermi's Golden Rule. It allows us to interpret Γ, the total decay rate, as the sum of the rates of transition to all lower states, which is entirely plausible. The effect of the finite lifetime of the upper state is to broaden the emission spectrum into the Lorentzian of width $\hbar\Gamma$ instead of an infinitely sharp delta function. This is *lifetime broadening*, though only for very short lifetime is the broadening observable without extreme experimental accuracy. For example, a lifetime of 1 ns corresponds to a wavenumber interval of 0.033 cm^{-1}, and 1 ps to 33 cm^{-1}.

The results in this section would appear to be incompatible with the use of a first-order approximation neglecting indirect transitions via third states. The method used allows the long-time limit to be taken because depletion of the initial state is included self-consistently. However, indirect transitions are still excluded.

14.7 Two-level system

The behaviour of a system consisting of only two coupled levels is particularly simple and provides insight into time-dependent phenomena generally. The two-level system also provides a useful model which is invoked in theories of the response of disordered systems. Furthermore, it is relevant to certain resonance and relaxation phenomena.

Consider a system with states $|1\rangle$ and $|2\rangle$ of energies E_1 and E_2 such that $E_1 - E_2 = 2\hbar\Delta$. The states may be degenerate, for example when they correspond to one or other of a pair of non-interacting molecules being in an excited state. The system is subject to a constant perturbation $\hat{V}$ with matrix elements such that $V_{11} = 0 = V_{22}$ (any non-zero diagonal elements can be subsumed in E_1 and E_2) and $V_{12} = \hbar V = V_{21}$, where for hermiticity V must be real. Then the state of the system can be written as

$$|v(t)\rangle = c_1(t)|1\rangle e^{-iE_1t/h} + c_2(t)|2\rangle e^{-iE_2t/\hbar}, \tag{14.101}$$

where from eq. (14.57) the coefficients satisfy the exact equations

$$dc_1/dt = -ic_2Ve^{2i\Delta t} \tag{14.102}$$

$$dc_2/dt = -ic_1Ve^{-2i\Delta t}. \tag{14.103}$$

These equations are separated by differentiating eq. (14.102) and substituting in eq. (14.103), and vice versa, to yield

$$d^2c_1/dt^2 - 2i\Delta dc_1/dt + V^2c_1 = 0 \tag{14.104}$$

$$d^2c_2/dt^2 + 2i\Delta dc_2/dt + V^2c_2 = 0. \tag{14.105}$$

The general solutions of these equations are

$$c_1(t) = (Ae^{i\Omega t} + B\,e^{-i\Omega t})e^{i\Delta t} \tag{14.106}$$

$$c_2(t) = (Ce^{-i\Omega t} + D\,e^{i\Omega t})e^{-i\Delta t}, \tag{14.107}$$

where $\Omega = (\Delta^2 + V^2)^{1/2}$. The coefficients A, B, C and D are determined by normalization and by the boundary conditions in particular cases.

Suppose that the system is prepared in state 1 at time zero, for example by exciting one of a pair of molecules. Then $c_1(0) = 1$ and $c_2(0) = 0$, so that $B = 1 - A$ and $D = -C$. These results are substituted back in (14.106) and

(14.107), when comparison of coefficients of $e^{\pm i\Omega t}$ yields $A = \frac{1}{2}(1 - \Delta/\Omega)$ and $C = \frac{1}{2}V/\Omega$; thence

$$c_1(t) = [\cos \Omega t - i(\Delta/\Omega) \sin \Omega t] e^{i\Delta t} \tag{14.108}$$

$$c_2(t) = [i(V/\Omega) \sin \Omega t] e^{-i\Delta t}. \tag{14.109}$$

The probability of finding the system in state 2 after time t is $|c_2(t)|^2$, which is also the probability $P(1 \rightarrow 2)$ that the system has undergone a transition from state 1 to state 2 after this time, i.e.

$$P(1 \rightarrow 2) = (V^2/\Omega^2) \sin^2 \Omega t \tag{14.110}$$

$$= [V^2/(V^2 + \Delta^2)] \sin^2 \Omega t. \tag{14.111}$$

The first thing to notice about the expressions (14.110) and (14.111) is that they are oscillatory in time. The probability $P(1 \rightarrow 2)$ oscillates between zero and V^2/Ω^2. When states 1 and 2 are degenerate, $\Delta = 0$ and $\Omega = V$, so that the system oscillates back and forth indefinitely between states 1 and 2, either of which is completely populated at different times. When the energy mismatch Δ is non-zero, the population of the initial state can never be depleted to zero, but still oscillates between Δ^2/Ω^2 and 1. The second point about the probability $P(1 \rightarrow 2)$ is that it is never a linear function of time: even at short times such that $\Omega t \ll 1$, $P(1 \rightarrow 2)$ varies as t^2, behaviour typical for a *coherent* process. In a finite system, the original state is always regained after a sufficiently long time.

Irreversibility is, however, obtained if the two-level system is coupled to a quasi-continuum of states which provides a density of final states, as noted in Section 14.4. This feature is utilized in a number of coherent phenomena in spectroscopy, including spin echoes in magnetic resonance and photon echoes in optical spectroscopy. These phenomena depend on turning the interaction V on or off at suitable multiples of the time interval $2\pi/\Omega$. This can regenerate an initially prepared state via the coherent properties of the system, yielding a pulse or echo of an intensity which reflects the incoherent irreversible processes which have occurred meanwhile.

Further insight into the transition process can be obtained by going back to the Hamiltonian $\hat{H}$ for the interacting two-level system. Its matrix elements in the basis of the non-interacting states $|1\rangle$ and $|2\rangle$ are

$$\{H_{ij}\} = \begin{pmatrix} E_1 & \hbar V \\ \hbar V & E_2 \end{pmatrix}. \tag{14.112}$$

The eigenvalues W are then found by solving the determinantal equation

$$\begin{vmatrix} E_1 - W & \hbar V \\ \hbar V & E_2 - W \end{vmatrix} = 0 \tag{14.113}$$

whence (cf. (14.36))

$$W_{\pm} = \tfrac{1}{2}(E_1 + E_2) \pm \hbar\Omega. \tag{14.114}$$

In the degenerate case, $W_{\pm} = E \pm \hbar V$, and the eigenvectors of $\hat{H}$ are found to be

$$|\pm\rangle = [|1\rangle \pm |2\rangle]/2^{1/2}. \tag{14.115}$$

If now c_1 and c_2 are evaluated from eqs (14.106) and (14.107) in the degenerate case from the initial condition $c_1(0) = 1/2^{1/2} = \pm c_2(0)$, it is found that the resulting state $|v(t)\rangle$ depends on time through a single factor $\exp[-i(E \pm \hbar V)t/\hbar]$. These symmetric and antisymmetric combinations of the non-interacting states are eigenstates of $\hat{H}$ and hence *pure states* with a simple time dependence. When the interacting system is prepared in state $|1\rangle$, it is a mixture of the two pure states and hence exhibits a composite time dependence instead of a single complex exponential corresponding to a single energy state.

Examples

14.1 A particle of mass m and charge q is confined to the x-axis by a potential which is zero for $0 \leqslant x \leqslant a$ and infinite elsewhere. The eigenvalues and normalized eigenfunctions are then

$$E_n^{(0)} = n^2h^2/8ma^2,$$

$$\psi_n^{(0)} = (2/a)^{1/2} \sin(n\pi x/a),$$

where $n = 1, 2, \ldots$. An electric field F is applied, producing the perturbation qFx. Show that the first-order correction to the ground-state energy is $\frac{1}{2}qFa$ and that the second-order correction is

$$-(512ma^4q^2F^2/\pi^4h^2) \sum_{\text{even } n}^{\infty} n^2/(n^2-1)^5.$$

For positive non-zero integers p,

$$\int_0^{\pi} y \cos py \, dy = 0, \qquad p \text{ even};$$
$$= -2/p^2, \qquad p \text{ odd}.$$

14.2 Calculate the first-order perturbation energy and the coefficients of the zeroth-order states when the electric field perturbation $-eFz$ is applied to the hydrogen atom in its degenerate $n = 2$ state. (Show that the only non-zero matrix element of z between the various $n = 2$ states is $\langle 2s|z|2p_z\rangle$; its value is $-3a$, where a is the Bohr radius.)

14.3 Deduce the pure states for the degenerate interacting two-level system directly from the coupled Schrödinger equations (14.102) and (14.103).

Further reading

P. W. Atkins, *Molecular Quantum Mechanics*, 2nd edn, Clarendon Press, Oxford, 1983.

J. L. Martin, *Basic Quantum Mechanics*, Clarendon Press, Oxford, 1981.

P. Landshoff and A. Metherell, *Simple Quantum Physics*, Cambridge University Press, 1979.
J. P. Lowe, *Quantum Chemistry*, Academic Press, New York, 1978.
I. N. Levine, *Quantum Chemistry*, 3rd edn, Allyn and Bacon, Boston, 1983.
G. Henderson, How a photon is created or absorbed, *J. Chem. Ed.*, **56** (1979) 631.
D. Shalitin, On the time-energy uncertainty relation, *Am. J. Phys.*, **52** (1984) 1111.

Chapter 15

Calculation of molecular properties

In Part B we have given a classical phenomenological description of the electromagnetic properties of molecules and molecular assemblies. For example, given that molecules may have permanent or induced dipole moments, we have shown how the optical and dielectric properties of molecular assemblies follow. In this and the following chapter, we treat the microscopic quantum-mechanical theories of the molecular properties. The present chapter concentrates on the electromagnetic properties introduced in Part B, while the next chapter deals with changes of quantum state and principles of spectroscopy.

15.1 Electric moments

The electric multipole moments were introduced in Chapter 8, where we saw how they determine the potential due to a charge distribution and the energy of a charge distribution in a field. Interactions between multipole moments also help to determine the local structure in molecular fluids and solids.

Expression for the multipole moments have already been given in terms of sums over a discrete distribution of charges (eqs (8.5)–(8.7)) and in terms of integrals over a continuous charge distribution (eqs (8.11)–(8.13)). For molecules and ions the total charge is usually known *a priori*. The calculated molecular state $|i\rangle$ is normally expressed in terms of the electronic wavefunction $\psi_i(\boldsymbol{r})$ such that the probability of finding an electron in a volume element $d\tau$ centred on the position $\boldsymbol{r}$ is $|\psi_i(\boldsymbol{r})|^2$. Since $|\psi_i(\boldsymbol{r})|^2$ is the probability density, the charge density is $-e|\psi_i(\boldsymbol{r})|^2$, where e is the proton charge and $-e$ therefore the electron charge. The nuclei also have discrete charges $Z_N e$ at the nuclear positions $\boldsymbol{r}_N$.

The molecular dipole moment is then given by a combination of the discrete and continuous nuclear and electronic contributions as

$$p_i = e\left[\sum_N z_N \boldsymbol{r}_N - \int \boldsymbol{r}\,|\psi_i(\boldsymbol{r})|^2\,d\tau\right]. \tag{15.1}$$

Alternatively the integral can be recognized as the scalar product $\langle i|\boldsymbol{r}|i\rangle$ giving the expectation value of $\boldsymbol{r}$ in the state $|i\rangle$, so that

$$p_i = e\left[\sum Z_N \boldsymbol{r}_N - \langle i|\boldsymbol{r}|i\rangle\right]. \tag{15.2}$$

A still more compact but less explicit expression is obtained by noting that $q\hat{\boldsymbol{R}}$ is the dipole moment operator, so that

$$p_i = \langle i|q\hat{\boldsymbol{R}}|i\rangle, \tag{15.3}$$

where it is to be understood that nuclear contributions are included. Equation (15.3) shows the observable dipole moment in state $|i\rangle$ as the expectation value of the corresponding operator in that state, in accordance with the general principles discussed in Section 11.3.

Higher electric multipole moments are given by formulae directly analogous to eqs (15.1)–(15.3). For example, the quadrupole moment can be written as

$$\mathbf{Q}_i = e\left[\sum_N Z_N \boldsymbol{r}_N \boldsymbol{r}_N - \langle i|\boldsymbol{rr}|i\rangle\right]. \tag{15.4}$$

Properties such as the electric multipole moments are described as 'one-electron' properties: they depend only on the positions of individual electrons separately, and not on the relative positions of pairs of electrons. Suites of computer programs for calculating molecular wavefunctions routinely include a package to calculate such one-electron properties.

15.2 Polarizability

Polarizabilities were treated in Sections 8.4 to 8.6. Here we shall develop the quantum-mechanical analogue of the model calculation in Section 8.6. An alternating electric field is applied to a molecule, and we seek that part of the molecular dipole moment which is oscillating at the same frequency and is proportional to the magnitude of the field. The constant of proportionality is the polarizability at the frequency in question.

The perturbation is, as in Section 13.4,

$$\hat{V}(t) = -q\hat{\boldsymbol{R}}\cdot\boldsymbol{E}(t), \tag{15.5}$$

where $q\hat{\boldsymbol{R}}$ is again the dipole operator. The electric field is supposed to oscillate at frequency ω, so that

$$\boldsymbol{E}(t) = \boldsymbol{E}_0 \cos \omega t = \tfrac{1}{2}\boldsymbol{E}_0(\mathrm{e}^{i\omega t} + \mathrm{e}^{-i\omega t}). \tag{15.6}$$

Now as we saw in the previous chapter, switching on such a perturbation will cause transitions between quantum states instead of leaving the molecule in its initial state $|k\rangle$. We therefore use the device of switching on the perturbation adiabatically (Section 14.5), with

$$\hat{V}'(t) = \hat{V}(t)\mathrm{e}^{\gamma t}, \tag{15.7}$$

where γ is arbitrarily small but positive. The perturbation can then be written as

$$\hat{V}'(t) = \tfrac{1}{2}\hat{V}(e^{i\omega t} + e^{-i\omega t})e^{\gamma t}, \tag{15.8}$$

where the magnitude $\hat{V}$ is given by

$$\hat{V} = -q\hat{\boldsymbol{R}}\cdot\boldsymbol{E}_0. \tag{15.9}$$

We now apply first-order time-dependent perturbation theory. The state of the molecule at time t is

$$|\nu(t)\rangle = |k,t\rangle + \sum_{m(\neq k)} c_m(t)\,|m,t\rangle, \tag{15.10}$$

where the states $|m\rangle$ are the eigenstates of the system in the absence of the electric field perturbation. The coefficients $c_m(t)$ are given by eq. (14.63), which takes the form

$$c_m(t) = (V_{mk}/2i\hbar)\int_{-\infty}^{t} (e^{i\omega t'} + e^{-i\omega t'})\, e^{\gamma t'} e^{i\omega_{mk}t'}\, dt'. \tag{15.11}$$

where $V_{mk} = \langle m|\hat{V}|k\rangle$ as usual. The switching-on factor $e^{\gamma t'}$ makes the integral vanish at its lower limit $t' \to -\infty$, leaving

$$c_m(t) = \frac{V_{mk}}{2i\hbar}\left[\frac{e^{i(\omega_{mk}+\omega)t}}{i(\omega_{mk}+\omega-i\gamma)} + \frac{e^{i(\omega_{mk}-\omega)t}}{i(\omega_{mk}-\omega-i\gamma)}\right]. \tag{15.12}$$

We choose the parameter γ so that it is much smaller than any of the quantities $\omega_{mk} \pm \omega$ and so that $\gamma t \ll 1$ for any realistic time of observation; the factors in γ can then be neglected in eq. (15.12), and

$$c_m(t) = -\frac{V_{mk}}{2\hbar}\left[\frac{e^{i(\omega_{mk}+\omega)t}}{\omega_{mk}+\omega} + \frac{e^{i(\omega_{mk}-\omega)t}}{\omega_{mk}-\omega}\right]. \tag{15.13}$$

The coefficients $c_m(t)$ now have only the applied frequency governing their time dependence, apart from the frequencies ω_{mk} governing the intrinsic time dependence of the states $|m\rangle$ and $|k\rangle$.

The dipole moment in the perturbed state $|\nu(t)\rangle$ is

$$\boldsymbol{p}_k(\boldsymbol{E}) = \langle \nu(t)|q\hat{\boldsymbol{R}}|\nu(t)\rangle. \tag{15.14}$$

We substitute for $|\nu(t)\rangle$ from eq. (15.10) in this equation, omitting the terms in $c_m(t)^2$ because these contribute to the non-linear response together with the higher-order terms omitted from the expansion (15.10):

$$\boldsymbol{p}_k(\boldsymbol{E}) = \langle k,t|q\hat{\boldsymbol{R}}|k,t\rangle + \sum_{m(\neq k)}[c_m(t)\langle k,t|q\hat{\boldsymbol{R}}|m,t\rangle + c_m^{*}(t)\langle m,t|q\hat{\boldsymbol{R}}|k,t\rangle]. \tag{15.15}$$

Recall from eq. (14.47) that $|k,t\rangle = |k\rangle e^{-iE_k t/\hbar}$. The first term in eq. (15.15) is then simply $\boldsymbol{p}_k$, the permanent dipole moment in state $|k\rangle$, independent of field. Hence eq. (15.15) can be written as

$$p_k(E) = p_k + \sum_{m(\neq k)} [c_m(t) p_{km} e^{i\omega_{km}t} + \text{c.c.}], \tag{15.16}$$

where the letters c.c. denote the complex conjugate of the other term in square brackets, and

$$p_{km} = \langle k \mid q\hat{R} \mid m \rangle \tag{15.17}$$

is a transition dipole matrix element. Notice that similarly V_{mk} can be written from eq. (15.9) as

$$V_{mk} = -p_{mk} \cdot E_0. \tag{15.18}$$

Collecting up these results, we can write the induced dipole moment as

$$p_k(E) - p_k = \sum_{m(\neq k)} \left\{ (p_{km} e^{i\omega_{km}t} \, p_{mk} \cdot E_0 / 2\hbar) \times \left[\frac{e^{i(\omega_{mk}+\omega)t}}{\omega_{mk}+\omega} + \frac{e^{i(\omega_{mk}-\omega)t}}{\omega_{mk}-\omega} \right] + \text{c.c.} \right\}. \tag{15.19}$$

Here the factors $e^{i\omega_{km}t}$ and $e^{i\omega_{mk}t}$ combine to give unity, and adding the complex conjugate makes each term proportional to $(e^{i\omega t} + e^{-i\omega t})$:

$$p_k(E) - p_k = \sum_{m(\neq k)} (p_{km} p_{mk} / \hbar) \cdot \tfrac{1}{2} E_0 (e^{i\omega t} + e^{-i\omega t}) \times \left[\frac{1}{\omega_{mk}+\omega} + \frac{1}{\omega_{mk}-\omega} \right] \tag{15.20}$$

$$= \alpha_k(\omega) \cdot E(t) \tag{15.21}$$

by definition. Here $E(t)$ comes from eq. (15.6), and the frequency-dependent polarizability is seen to be given by

$$\alpha_k(\omega) = \frac{2}{\hbar} \sum_{m(\neq k)} \frac{\omega_{mk} p_{km} p_{mk}}{\omega_{mk}^2 - \omega^2}. \tag{15.22}$$

The polarizability in state $|k\rangle$ is thus a sum of terms for all states $|m\rangle$ other than $|k\rangle$. The contributions depend on the transition dipole moments p_{mk} and on the energy difference $\hbar\omega_{mk}$. States having energies E_m close to $E_k \pm \hbar\omega$ make the largest contribution to the polarizability provided their transition moments are not too small. As the frequency ω tends to infinity, α tends to zero. The polarizability is significant only so long as ω remains not too large in comparison with molecular electronic transition frequencies ω_{mk}, so that the inertia of the electrons is not too large for them to respond. As the frequency ω tends to zero, we get the static polarizability

$$\alpha_k = \alpha_k(0) = \frac{2}{\hbar} \sum_{m(\neq k)} \frac{p_{km} p_{mk}}{\omega_{mk}}. \tag{15.23}$$

In the ground state $E_k < E_m$ for all m so that $\omega_{mk} > 0$, and since $p_{mk} = p_{km}^*$ it can be seen that $\boldsymbol{\alpha}$ is always positive for the ground state. This means that the energy $-\frac{1}{2}\alpha E^2$ is always negative and hence stabilizing: the ground state will only distort on polarization in such a way as to lower the energy of polarization. Excited states, on the other hand, can have negative polarizabilities if the consequent destabilizing effect of the polarization energy is outweighed by the stabilizing effect of mixing in a contribution from the ground state.

For frequencies ω which equal one of the transition frequencies ω_{mk}, the polarizability given by eq. (15.22) diverges. In this case, however, the effect of the finite lifetime of excited states becomes important. As we saw in Section 14.6, when lifetime effects are included one has to replace ω_{mk} by $\omega_{mk} + \frac{1}{2}i\Gamma_{mk}$. Then as $\omega \to \pm\,\omega_{mk}$, $\boldsymbol{\alpha}_k(\omega)$ becomes a Lorentzian function with a maximum of $2\omega_{mk}\,|\,p_{km}p_{mk}\,|/(\frac{1}{2}\hbar\Gamma_{mk})^2$. This is the same functional form as obtained for a classical oscillator with a phenomenological damping, eq. (8.44). The frequency dependence of the polarizability near each transition frequency then looks like that shown for the refractive index in Fig. 9.1.

The polarizability given by eq. (15.22) refers to a particular set of axes O xyz. One might also require the mean molecular polarizability in a gas of freely-rotating molecules. This is obtained by writing

$$\bar{\alpha} = \tfrac{1}{3}(\alpha_{xx} + \alpha_{yy} + \alpha_{zz}) \tag{15.24}$$

$$= \frac{2}{3\hbar} \sum_{m(\neq k)} \frac{\omega_{mk}\,|\,p_{km}\,|^2}{\omega_{mk}^2 - \omega^2}. \tag{15.25}$$

Contributions from α_{xy} and so on average to zero because to each contribution there corresponds an equal and opposite one of equal probability.

15.3 Oscillator strength

As eq. (15.22) shows, the quantum-mechanical expression for the polarizability consists of a sum of terms resembling those for the classical driven oscillator in eq. (8.43). The oscillator frequency ω_0 corresponds to a transition frequency ω_{mk}. The comparison with the classical result can be carried further by introducing dimensionless *oscillator strengths* defined by

$$f_{mk} = (2m_e\omega_{mk}/\hbar)\,|\,\langle m\,|\,\hat{\boldsymbol{R}}\,|\,k\rangle\,|^2, \tag{15.26}$$

where m_e is the electron mass. Substitution in eq. (15.22), taken for simplicity to be a scalar, yields

$$\alpha_k(\omega) = \sum_{m(\neq k)} \frac{f_{mk}\,q^2/m_e}{\omega_{mk}^2 - \omega^2}. \tag{15.27}$$

Each term in the polarizability is now exactly of the ideal electron oscillator form except for the oscillator strength which measures the strength or effectiveness of the transition compared with that of the ideal oscillating electron.

The oscillator strengths obey the *sum rule*

$$\sum_m f_{mk} = Z, \tag{15.28}$$

where Z is the total number of electrons. This sum rule is associated with Thomas, Reiche and Kuhn, and originates in a relationship between position and momentum matrix elements. A molecular electronic Hamiltonian is typically of the form (for fixed nuclei)

$$\hat{H}_0 = \sum_j \hat{P}_j^2/2m_e + V(\hat{\boldsymbol{R}}_1,\hat{\boldsymbol{R}}_2, \ldots), \tag{15.29}$$

where $\hat{\boldsymbol{P}}_j$ and $\hat{\boldsymbol{R}}_j$ are the momentum and position operators for electron j. These operators commute for different electrons, and different components commute for a given electron, while the same components for a given electron satisfy

$$[\hat{X}_j,\hat{P}_{xj}] = i\hbar. \tag{15.30}$$

It then follows that for each electron

$$[\hat{\boldsymbol{R}}_j,\hat{H}_0] = (i\hbar/m_e)\hat{\boldsymbol{P}}_j. \tag{15.31}$$

Equating matrix elements of both sides of eq. (15.31) yields

$$\langle m \,|\, \hat{\boldsymbol{R}}_j\hat{H}_0 \,|\, k\rangle - \langle m \,|\, \hat{H}_0\hat{\boldsymbol{R}}_j \,|\, k\rangle = (i\hbar/m_e)\langle m \,|\, \hat{\boldsymbol{P}}_j \,|\, k\rangle. \tag{15.32}$$

But since $\hat{H}_0 \,|\, k\rangle = E_k \,|\, k\rangle$, we have

$$\langle m \,|\, \hat{\boldsymbol{R}}_j\hat{H}_0 \,|\, k\rangle = E_k\langle m \,|\, \hat{\boldsymbol{R}}_j \,|\, k\rangle, \tag{15.33}$$

and because $\hat{H}_0$ is also Hermitian

$$\langle m \,|\, \hat{H}_0\hat{\boldsymbol{R}}_j \,|\, k\rangle = \langle k \,|\, \hat{\boldsymbol{R}}_j\hat{H}_0 \,|\, m\rangle^* \tag{15.34}$$

$$= E_m\langle m \,|\, \hat{\boldsymbol{R}}_j \,|\, k\rangle, \tag{15.35}$$

so that

$$(E_k - E_m)\langle m \,|\, \hat{\boldsymbol{R}}_j \,|\, k\rangle = (i\hbar/m_e)\langle m \,|\, \hat{\boldsymbol{P}}_j \,|\, k\rangle. \tag{15.36}$$

Hence, finally we deduce that

$$\langle m \,|\, \hat{\boldsymbol{R}}_j \,|\, k\rangle = \langle m \,|\, \hat{\boldsymbol{P}}_j \,|\, k\rangle / i m_e \omega_{mk} \tag{15.37}$$

and by summing over all electrons j that

$$\langle m \,|\, \hat{\boldsymbol{R}} \,|\, k\rangle = \langle m \,|\, \hat{\boldsymbol{P}} \,|\, k\rangle / i m_e \omega_{mk} \tag{15.38}$$

$$\langle k \,|\, \hat{\boldsymbol{R}} \,|\, m\rangle = -\langle k \,|\, \hat{\boldsymbol{P}} \,|\, m\rangle / i m_e \omega_{mk}. \tag{15.39}$$

The oscillator strength is now written as the sum of two identical terms:

$$f_{mk} = (m_e\omega_{mk}/\hbar)[\langle k \,|\, \hat{\boldsymbol{R}} \,|\, m\rangle\langle m \,|\, \hat{\boldsymbol{R}} \,|\, k\rangle + \langle k \,|\, \hat{\boldsymbol{R}} \,|\, m\rangle\langle m \,|\, \hat{\boldsymbol{R}} \,|\, k\rangle]. \tag{15.40}$$

We use eq. (15.38) to replace the second factor in the first term in square brackets and eq. (15.39) to replace the first factor in the second term. A factor

$m_e\omega_{mk}$ cancels, leaving

$$f_{mk} = (1/i\hbar)[\langle k|\hat{\boldsymbol{R}}|m\rangle\langle m|\hat{\boldsymbol{P}}|k\rangle - \langle k|\hat{\boldsymbol{P}}|m\rangle\langle m|\hat{\boldsymbol{R}}|k\rangle]. \quad (15.41)$$

On summation over all states $|m\rangle$, each product of matrix elements gives an element of the product matrix:

$$\sum_m f_{mk} = (1/i\hbar)[\langle k|\hat{\boldsymbol{R}}\hat{\boldsymbol{P}}|k\rangle - \langle k|\hat{\boldsymbol{P}}\hat{\boldsymbol{R}}|k\rangle] \quad (15.42)$$

$$= (1/i\hbar)\langle k|\hat{\boldsymbol{R}}\hat{\boldsymbol{P}} - \hat{\boldsymbol{P}}\hat{\boldsymbol{R}}|k\rangle \quad (15.43)$$

$$= (1/i\hbar)\langle k|\,[\hat{\boldsymbol{R}},\hat{\boldsymbol{P}}]\,|k\rangle. \quad (15.44)$$

From the discussion of the position and momentum operators before eq. (15.30), it follows that the commutator $[\hat{\boldsymbol{R}},\hat{\boldsymbol{P}}]$ is the sum of $[\hat{\boldsymbol{R}}_j,\hat{\boldsymbol{P}}_j]$ over all electrons j, and that each individual term is $i\hbar$. Thus

$$\sum_m f_{mk} = \sum_{j=1}^{Z} 1 = Z \quad (15.45)$$

as already asserted.

The sum rule is of use in indicating what contributions there are to the polarizability. It will be seen in Section 16.2 that the oscillator strength also governs spectral intensities, so that the sum rule can show whether intense transitions must be sought in unexplored regions of the spectrum. The sum rule is also used in an approximate expression for the polarizability obtained using the average denominator approximation of Section 14.1. We set $\omega_{mk} \approx \Omega_k$ for all m, so that from eq. (15.27)

$$\alpha_k(\omega) = \frac{q^2/m_e}{\Omega_k^2 - \omega^2} \sum_{m(\neq k)} f_{mk}. \quad (15.46)$$

The summation is now given by the sum rule as Z, except that the term f_{kk} has to be excluded. From eq. (15.26) it can be seen that f_{kk} is zero because of the factor ω_{mk}, so that we obtain

$$\alpha_k(\omega) \approx \frac{Zq^2/m_e}{\Omega_k^2 - \omega^2}. \quad (15.47)$$

This expression provides some theoretical basis for the trend noted in Section 8.6 for polarizability to increase with the number of electrons in a molecule.

The oscillator strengths for a one-dimensional harmonic oscillator of frequency ω_0 follow from the results

$$|\langle m|\hat{X}|k\rangle|^2 = (\hbar/2m\omega_0)k \qquad (m = k-1) \quad (15.48)$$

$$= (\hbar/2m\omega_0)(k+1) \qquad (m = k+1); \quad (15.49)$$

all matrix elements of the position operator $\hat{X}$ between non-adjacent states are zero. The non-zero oscillator strengths are then

$$f_{k-1,k} = -k \quad (15.50)$$

$$f_{k+1,k} = k+1, \quad (15.51)$$

where the minus sign in $f_{k-1,k}$ follows from the negative transition frequency $\omega_{k-1,k} = -\omega_0$. The polarizability of the harmonic oscillator in the state k becomes

$$\alpha_k = (q^2/m_e)\left[\frac{-k}{(-\omega_0)^2 - \omega^2} + \frac{k+1}{\omega_0^2 - \omega^2}\right] \tag{15.52}$$

$$= \frac{q^2/m_e}{\omega_0^2 - \omega^2} \tag{15.53}$$

This result is independent of k and agrees with the simple classical result in eq. (8.43). The sum rule for the harmonic oscillator strengths is readily verified as

$$\sum_m f_{mk} = -k + k + 1 = 1.$$

15.4 Magnetic moment

The main features of the magnetic moment have been dealt with in Section 10.1. Excluding nuclear contributions, the magnetic moment is simply the expectation value of the corresponding operator $\hat{\boldsymbol{M}}$ in the state in question. Here $\hat{\boldsymbol{M}}$ is

$$\hat{\boldsymbol{M}} = -(e/2m_e)(\hat{\boldsymbol{L}} + g_e\hat{\boldsymbol{S}}) \tag{15.54}$$

$$\approx -(e/2m_e)(\hat{\boldsymbol{L}} + 2\hat{\boldsymbol{S}}), \tag{15.55}$$

where we recall that g_e is the electron g-value 2.0023... and $-e/2m_e$ is the magnetogyric ratio γ. The problem in determining magnetic moments is that they usually refer to particular energy eigenstates which need not have a definite orbital angular momentum. A further complication is that spin and orbital angular momentum couple to affect the total energy.

In atoms and ions, the spherical average symmetry allows the electronic states to be expressed approximately as a superposition of orbitals of definite angular momentum—s, p, d, etc. When spin-orbit coupling is weak, the orbital angular moment $\hat{\boldsymbol{l}}$ for all the electrons are combined to give a total angular momentum $\hat{\boldsymbol{L}}$, and the spins $\hat{\boldsymbol{s}}$ are combined to give a total $\hat{\boldsymbol{S}}$; $\hat{\boldsymbol{L}}$ and $\hat{\boldsymbol{S}}$ are then combined to give a total angular momentum $\hat{\boldsymbol{J}}$. This is Russell–Saunders or LS-coupling. When spin-orbit coupling is strong, $\hat{\boldsymbol{l}}$ and $\hat{\boldsymbol{s}}$ are combined to give a total angular momentum $\hat{\boldsymbol{j}}$ for each electron, and the individual $\hat{\boldsymbol{j}}$ are combined to give $\hat{\boldsymbol{J}}$ again. This is jj-coupling, which applies mainly to the heavier atoms where spin-orbit coupling is strongest.

In diatomic molecules, the energy eigenstates have definite orbital angular momentum only for motion about the internuclear axis (i.e. definite components of angular momentum along the axis). These are the Σ, Π, Δ, etc. states analogous to atomic s, p, d, etc. Finally, in polyatomic molecules the energy eigenstates in general have no definite orbital angular momentum.

One useful explicit result can be derived for atoms and ions under

Russell–Saunders coupling. The interaction with a magnetic induction $\boldsymbol{B}$ is described by the operator in eq. (13.40), which for small $\boldsymbol{B}$ reduces to

$$\hat{H}_f = -\hat{\boldsymbol{M}}\cdot\boldsymbol{B} \tag{15.56}$$

$$\approx (e/2m_e)(\hat{\boldsymbol{L}} + 2\hat{\boldsymbol{S}})\cdot\boldsymbol{B}. \tag{15.57}$$

The scalar products $\hat{\boldsymbol{L}}\cdot\boldsymbol{B}$ and $\hat{\boldsymbol{S}}\cdot\boldsymbol{B}$ can be expressed in terms of the components of the orbital and spin angular momentum operators parallel to $\boldsymbol{B}$ (taken to lie along the z direction). However, these components are not quantized separately: only their sum is quantized, because it is the z component of the total angular momentum operator $\hat{\boldsymbol{J}} = \hat{\boldsymbol{L}} + \hat{\boldsymbol{S}}$, which is parallel to the magnetic moment operator only when $S = 0$, i.e. in singlet states. It would therefore be convenient if eq. (15.57) could be expressed in terms of $\hat{\boldsymbol{J}}\cdot\boldsymbol{B}$ and the quantum numbers J, L and S.

Now $\hat{H}_f$ can be written as

$$\hat{H}_f = -\gamma(\hat{\boldsymbol{J}}\cdot\boldsymbol{B} + \hat{\boldsymbol{S}}\cdot\boldsymbol{B}), \tag{15.58}$$

so that the problem is to express $\hat{\boldsymbol{S}}\cdot\boldsymbol{B}$ as a multiple of $\hat{\boldsymbol{J}}\cdot\boldsymbol{B}$. This can be done since the components of $\hat{\boldsymbol{S}}$ perpendicular to $\hat{\boldsymbol{J}}$ are indeterminate, leaving the result

$$\hat{\boldsymbol{S}}\cdot\boldsymbol{B} = (\hat{\boldsymbol{S}}\cdot\hat{\boldsymbol{J}}/\hat{J}^2)(\hat{\boldsymbol{J}}\cdot\boldsymbol{B}). \tag{15.59}$$

The requisite scalar product is therefore obtainable, since

$$\hat{\boldsymbol{S}}\cdot\hat{\boldsymbol{J}} = \tfrac{1}{2}[\hat{J}^2 + \hat{S}^2 - (\hat{J} - \hat{S})^2] \tag{15.60}$$

$$= \tfrac{1}{2}(\hat{J}^2 + \hat{S}^2 - \hat{L}^2), \tag{15.61}$$

where the eigenvalues of all the squared operators are determinate as $\hbar^2 \times J(J+1)$, etc. We therefore have

$$\hat{H}_f = -\gamma g\hat{\boldsymbol{J}}\cdot\boldsymbol{B}, \tag{15.62}$$

where g is the *Landé g-factor* given by

$$g = 1 + [J(J+1) + S(S+1) - L(L+1)]/2J(J+1). \tag{15.63}$$

As can be seen, when $S = 0$ and hence $J = L$, the g-factor reduces to unity, while when $L = 0$ and hence $J = S$, the g-factor reduces to two.

15.5 Magnetizability

The magnetizability $\varkappa$ can be calculated in much the same way as the polarizability. However, the calculation includes some complicating features, and so for simplicity we shall concentrate on the static magnetizability, merely stating without proof the modification required in a periodic magnetic induction. It is then convenient to calculate the magnetizability not in terms of the magnetic moment induced by the magnetic induction but in terms of the energy change produced by the magnetic induction. From eq. (10.8) we know that the change in the energy to second order in the magnetic induction is

$$W = -\tfrac{1}{2}\varkappa B^2, \tag{15.64}$$

where an isotropic magnetizability is assumed. We use time-independent perturbation theory to calculate W and hence $\varkappa$, incidentally illustrating the usefulness of perturbation theory in this area.

The perturbation is given by eq. (13.40) as

$$\hat{V} = -\hat{M}_z B + (q^2B^2/8m)(\hat{X}^2 + \hat{Y}^2) \tag{15.65}$$

where $\hat{H}_f$ is now written as $\hat{V}$ to conform with the notation in Chapter 14, since no confusion with the scalar potential will arise here. We see that $\hat{V}$ can be separated into a part $\hat{V}_1$ proportional to B and a part $\hat{V}_2$ proportional to B^2. An energy change which is second-order in B thus comes from the second-order term in $\hat{V}_1$ and the first-order term in $\hat{V}_2$. In the state $|k\rangle$ we therefore have to second order in B

$$W_k = \langle k|\hat{V}_2|k\rangle + \sum_{m(\neq k)} \frac{|\langle m|\hat{V}_1|k\rangle|^2}{E_k - E_m}, \tag{15.66}$$

where we have used the perturbation theory results of eqs (14.13) and (14.26). Using the explicit forms of $\hat{V}_1$ and $\hat{V}_2$ yields

$$W_k = (q^2B^2/8m)\langle k|\hat{X}^2 + \hat{Y}^2|k\rangle + B^2 \sum_{m(\neq k)} \frac{|\langle m|\hat{M}_z|k\rangle|^2}{E_k - E_m}. \tag{15.67}$$

with the expected dependence on B^2.

In most cases it is the ground state $|0\rangle$ in which we are interested, and then

$$W_0 = \frac{q^2B^2}{8m}\langle 0|\hat{X}^2 + \hat{Y}^2|0\rangle - B^2 \sum_{m(\neq 0)} \frac{|\langle m|\hat{M}_z|0\rangle|^2}{\Delta E_m} \tag{15.68}$$

where $\Delta E_m = E_m - E_0$ is the excitation energy. It can be seen that in this equation the first term on the right-hand side is intrinsically positive and the second negative. The total magnetizability is therefore a sum of diamagnetic and paramagnetic terms, $\varkappa = \varkappa_d + \varkappa_p$. Comparison with eq. (15.64) gives these terms as

$$\varkappa_d = -(q^2/4m)\langle 0|\hat{X}^2 + \hat{Y}^2|0\rangle \tag{15.69}$$

$$\varkappa_p = 2\sum_{m(\neq 0)} |\langle m|\hat{M}_z|0\rangle|^2/\Delta E_m. \tag{15.70}$$

The diamagnetic term has no electric analogue, since it arises from the B^2 term in the interaction $\hat{V}$. In a fluid sample, the expectation value of $\hat{X}^2$ is the same as that of $\hat{Y}^2$ and $\hat{Z}^2$, so that each is equal to a third of the expectation value of $\hat{R}^2$, and then

$$\varkappa_d = -(e^2/6m)\langle 0|\hat{R}^2|0\rangle. \tag{15.71}$$

Because $\varkappa_d$ depends only on the molecular ground state, it is relatively easy to evaluate. It is also quite well reproduced by a set of atomic contributions transferable between different molecules. This additivity leads to *Pascal's rules* for structural contributions to $\varkappa_d$. These work well except in aromatic compounds where ring currents destroy simple additivity.

The orbital paramagnetic term is usually much smaller than the diamagnetic term: dominant paramagnetic behaviour comes from a non-zero spin S as discussed in Chapter 10. For a fluid sample, averaging again introduces a factor of one-third, and expressing $\hat{\boldsymbol{M}}$ in terms of the orbital angular momentum operator $\hat{\boldsymbol{L}}$ gives

$$\varkappa_p = (e^2/6m^2) \sum_{m(\neq 0)} |L_{m0}|^2/\Delta E_m, \tag{15.72}$$

where $L_{m0} = \langle m | \hat{L} | 0 \rangle$. This governs the *temperature-independent paramagnetism* or TIP, in contrast to the temperature-dependent spin paramagnetism. The TIP is the magnetic analogue of the polarizability, depending on the distortion of the molecular states. When the energy eigenstates $|m\rangle$ are also eigenstates of angular momentum, the matrix elements L_{m0} are proportional to the overlap integrals $\langle m | 0 \rangle$, which are zero. Thus for light atoms the TIP is usually zero.

In an oscillating magnetic induction of frequency ω, there are still diamagnetic and paramagnetic contributions. The diamagnetic contribution is unchanged, but the paramagnetic magnetizability becomes

$$\varkappa_p(\omega) = 2 \sum_{m(\neq 0)} \frac{\Delta E_m |\langle m | \hat{M}_z | 0 \rangle|^2}{\Delta E_m^2 - (\hbar\omega)^2}. \tag{15.73}$$

This is exactly analogous to eq. (15.22) for $\alpha(\omega)$. The TIP is sometimes called the high-frequency component of paramagnetism. Although it tends to zero at high frequencies, it does so only when the inertia of the electrons becomes too great, whereas the orientation which produces the spin paramagnetism tends to zero at much lower frequencies when the inertia of the nuclei becomes too great. This is directly analogous to the 'freezing out' of orientation polarization at high frequencies, leaving only distortion polarization (Section 9.4).

Finally it should be noted that the separation of the magnetizability into paramagnetic and diamagnetic contributions depends on the choice of gauge for the vector potential. The present choice of the Coulomb gauge is the conventional one. Other choices give a different separation, so that the paramagnetic and diamagnetic contributions are not independent of gauge and hence are not proper measurable quantities. The total magnetizability is, of course, measurable and can be shown to be independent of the choice of gauge.

15.6 Optical activity and rotational strengths

In Section 9.6 we saw how optical activity ensues when the induced electric dipole moment depends not only on the electric field but also on the spatial variation of the field through a term $\alpha'(\nabla \times E)$. The molar rotation $[\phi]$ is then proportional to α', which can be interpreted as depending on the imaginary part of the total complex polarizability. An alternative interpretation, which allows earlier results in this chapter to be adapted to calculate α', is to recall the Maxwell equation

$$\nabla \times \boldsymbol{E} = -\partial \boldsymbol{B}/\partial t. \tag{15.74}$$

Hence α' can be regarded as giving the *electric* dipole moment induced by a time-dependent *magnetic induction.*

The calculation of α' proceeds as in Section 15.2 via the expectation value of the dipole operator $q\hat{\boldsymbol{R}}$ in the perturbed state, except that the perturbation operator is now the term $V_1 = -\hat{\boldsymbol{M}}\cdot\boldsymbol{B}$ from eq. (15.56). Because the term V_2 is quadratic in $\boldsymbol{B}$, it cannot contribute to the linear response. The induced dipole in the ground state is then given by

$$\boldsymbol{p}_0(\boldsymbol{B}) = \boldsymbol{p}_0 + \sum_{k(\neq 0)} \left[c_k(t)\boldsymbol{p}_{0k}\mathrm{e}^{i\omega_{0k}t} + \text{c.c.} \right], \tag{15.75}$$

where the coefficients $c_k(t)$ depend on the perturbation matrix element (cf. eq. (15.18))

$$V_{k0} = -\boldsymbol{m}_{k0}\cdot\boldsymbol{B}_0, \tag{15.76}$$

with $\boldsymbol{m}_{k0} = \langle k|\hat{\boldsymbol{M}}|0\rangle$ the magnetic dipole transition matrix element and $\boldsymbol{B}_0$ the amplitude of $\boldsymbol{B}$ such that $\boldsymbol{B} = \frac{1}{2}\boldsymbol{B}_0(\mathrm{e}^{i\omega t} + \mathrm{e}^{-i\omega t})$. Then

$$\boldsymbol{p}_0(\boldsymbol{B}) - \boldsymbol{p}_0 = \frac{1}{2\hbar}\sum_{k(\neq 0)} \left[\boldsymbol{p}_{0k}\boldsymbol{m}_{k0}\cdot\boldsymbol{B}_0\left(\frac{\mathrm{e}^{i\omega t}}{\omega_{k0}+\omega} + \frac{\mathrm{e}^{-i\omega t}}{\omega_{k0}-\omega}\right) + \text{c.c.} \right]. \tag{15.77}$$

In evaluating the complex conjugate we need the results that $\boldsymbol{p}_{0k}$ is real because the dipole operator $q\hat{\boldsymbol{R}}$ is real, whereas $\boldsymbol{m}_{k0}$ is imaginary because the magnetic dipole operator is proportional to $\hat{\boldsymbol{L}} = \hat{\boldsymbol{R}} \times \hat{\boldsymbol{P}}$ where $\hat{\boldsymbol{P}} = -i\hbar\nabla$ and hence is imaginary. As a result, $\boldsymbol{p}_{0k}^* = \boldsymbol{p}_{0k}$ but $\boldsymbol{m}_{k0}^* = -\boldsymbol{m}_{k0}$, so that eq. (15.77) leads to

$$\boldsymbol{p}_0(\boldsymbol{B}) - \boldsymbol{p}_0 = -\left(\frac{\omega}{\hbar}\right)\sum_{k(\neq 0)} \boldsymbol{p}_{0k}\boldsymbol{m}_{k0}\cdot\boldsymbol{B}_0\frac{(\mathrm{e}^{i\omega t} - \mathrm{e}^{-i\omega t})}{\omega_{k0}^2 - \omega^2}. \tag{15.78}$$

Now from eqs (9.49) and (15.74) it can be seen that α' is the coefficient of proportionality between $\boldsymbol{p}_0(\boldsymbol{B}) - \boldsymbol{p}_0$ and $-\partial\boldsymbol{B}/\partial t$, and here

$$\partial\boldsymbol{B}/\partial t = \tfrac{1}{2}i\omega\boldsymbol{B}_0(\mathrm{e}^{i\omega t} - \mathrm{e}^{-i\omega t}). \tag{15.79}$$

Comparison with eq. (15.78) then yields the result

$$\boldsymbol{\alpha}' = (2/i\hbar)\sum_{k(\neq 0)} \frac{\boldsymbol{p}_{0k}\boldsymbol{m}_{k0}}{\omega_{k0}^2 - \omega^2}. \tag{15.80}$$

This is real because $\boldsymbol{m}_{k0}$ is imaginary, and so can be written alternatively as

$$\boldsymbol{\alpha}' = (2/\hbar)\mathrm{Im}\sum_{k(\neq 0)} \frac{\boldsymbol{p}_{0k}\boldsymbol{m}_{k0}}{\omega_{k0}^2 - \omega^2}. \tag{15.81}$$

The form of $\boldsymbol{\alpha}'$ shows that this coefficient also governs the *magnetic* dipole moment induced by a time-dependent *electric* field. This would be given by the same expression, but with $\boldsymbol{p}_{0k}\boldsymbol{m}_{k0}$ replaced by $\boldsymbol{m}_{0k}\boldsymbol{p}_{k0}$. However, since the two dipole operators are Hermitian, this replacement yields the complex conjugate of the original expression, which being real is unchanged.

The quantity

$$R_{kl} = \mathrm{Im}\, \boldsymbol{p}_{kl} \cdot \boldsymbol{m}_{lk} \tag{15.82}$$

is the *rotational strength* of the transition $k \to l$, analogous to the oscillator strength. It measures the strength of the contribution of a given transition to the optical rotation. It is defined in terms of the scalar product of the two vector matrix elements, because this determines the mean response in a gas of freely-rotating molecules, namely

$$\bar{\alpha}' = (2/3\hbar) \sum_{k(\neq 0)} R_{0k}/(\omega_{k0}^2 - \omega^2). \tag{15.83}$$

The factor of 1/3 is analogous to that in eq. (15.25) for the mean polarizability $\bar{\alpha}$.

The rotational strengths determine whether or not a system can show optical activity. Only systems which are *dissymmetric,* i.e. have neither a plane nor a centre of symmetry, can have non-zero rotational strengths. Note incidentally that a helix is dissymmetric but does have a well-defined symmetry, so that the frequent use of the word 'asymmetric' in the context of optical activity is too restrictive in its implications. On inversion, the operators $\hat{\boldsymbol{R}}$ and $\hat{\boldsymbol{P}}$ change sign and hence $\hat{\boldsymbol{L}}$ does not, whence $\boldsymbol{p}_{0k}$ changes sign and $\boldsymbol{m}_{k0}$ does not, so that the rotational strength R_{0k} changes sign. On reflection, only the components of $\hat{\boldsymbol{R}}$ and $\hat{\boldsymbol{P}}$ perpendicular to the mirror plane change sign and hence only the components of $\hat{\boldsymbol{L}}$ in the plane change sign, so that corresponding components of $\boldsymbol{p}_{0k}$ and $\boldsymbol{m}_{k0}$ are of opposite sign, and the rotational strength again changes sign. If either inversion or reflection is a symmetry operation for the system in question, it must leave R_{0k} unchanged, which implies that R_{0k} must be zero.

For a dissymmetric molecule, reflection in a mirror plane yields not the same molecule but its enantiomer. The rotational strengths can then be non-zero, but since R_{0k} changes sign on reflection, the rotational strengths of enantiomeric pairs of molecules must be equal and opposite. Their values of α' and hence by eq. (9.62) of the molar rotation $[\phi]$ must therefore also be equal and opposite. This agrees with experimental observation, and accounts for the fact that racemic (i.e. equimolar) mixtures of enantiomers show zero optical rotation.

The helix provides a simple model to help visualize how optical activity occurs. In a helix, such as a solenoidal winding, translation along the axis is associated with rotation of definite handedness about the axis. Suppose the light is propagating at right angles to the axis of the helix. Two limiting cases occur, as illustrated in Fig. 15.1: either $\boldsymbol{E}$ or $\boldsymbol{B}$ can be parallel to the axis of the helix. In the former case (a), the oscillating electric field drives electrons

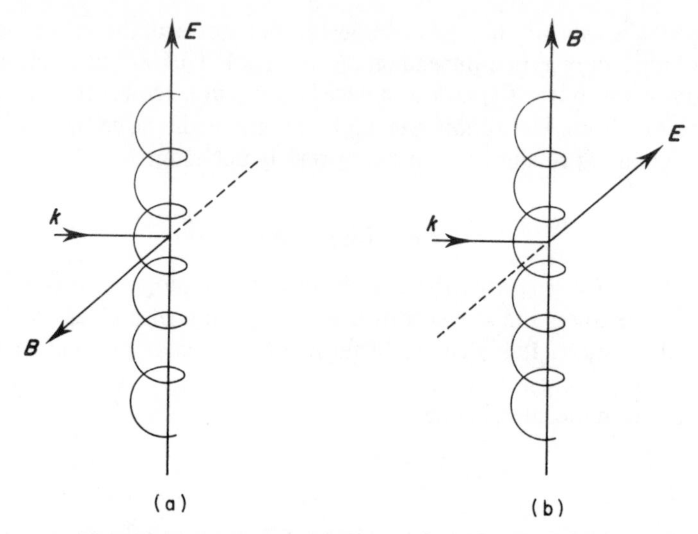

Fig. 15.1 Optical rotation in a helix. The light progagates with its wavevector $\boldsymbol{k}$ perpendicular to the axis of the helix. In case (a) the electric field vector is parallel to the axis of the helix, and in case (b) the magnetic induction vector is parallel to the axis; the third vector is perpendicular to the other two as shown

up and down the helix. This produces not only an oscillating electric dipole moment along the axis, but also an oscillating magnetic dipole along the axis, by virtue of the circular component of motion along the axis. The sense of the helix determines that of the circular motion of the electrons and hence whether the magnetic moment is parallel or antiparallel to the electric moment. In the latter case (b), the oscillating magnetic induction drives electrons around the helix and back. This produces not only an oscillating magnetic dipole moment along the axis, but also an electric dipole moment along the axis by virtue of the axial component of the circular motion about the axis. The sense of the helix again determines whether the electric and magnetic moments are parallel or antiparallel. Other dissymmetric molecules have a handedness which leads similarly to a linking of translational and rotational motion and hence to optical activity.

The frequency dependence of optical rotation follows from eqs (9.62) and (15.83), which give the molar rotation as

$$[\phi] = -(2N_{\mathrm{A}}\omega^2\mu_0/3\hbar)\sum_{k(\neq 0)} R_{0k}/(\omega_{k0}^2 - \omega^2). \tag{15.84}$$

At low frequencies, α' tends to a constant independent of frequency, but $[\phi]$ becomes proportional to ω^2 and so tends to zero. At high frequencies ($\omega \gg \omega_{k0}$

for all k), the molar rotation tends to

$$[\phi]_\infty = (2N_A\mu_0/3\hbar) \sum_{k(\neq 0)} R_{0k}. \qquad (15.85)$$

The unrestricted sum of rotational strengths over all states l is

$$\sum_l R_{kl} = \mathrm{Im} \sum_l \boldsymbol{p}_{kl} \cdot \boldsymbol{m}_{lk}. \qquad (15.86)$$

This can be recognized as the kk matrix element of the product of the electric and magnetic transition dipole moment operators. The electric dipole operator is proportional to $\hat{\boldsymbol{R}}$, while the magnetic dipole operator is proportional to $\hat{\boldsymbol{R}} \times \hat{\boldsymbol{P}}$, as already noted. However, $\hat{\boldsymbol{R}} \times \hat{\boldsymbol{P}}$ is necessarily perpendicular to $\hat{\boldsymbol{R}}$ and hence has zero scalar product with it, so that we have the sum rule

$$\sum_l R_{kl} = 0. \qquad (15.87)$$

The restricted sum in eq. (15.85) is therefore $-R_{00} = -\mathrm{Im}\, \boldsymbol{p}_0 \cdot \boldsymbol{m}_0$, where $\boldsymbol{p}_0$ and $\boldsymbol{m}_0$ are the permanent electric and magnetic dipole monents in the ground state. Since these are real quantities, they have zero imaginary part, and $R_{00} = 0$. Consequently the high-frequency molar rotation is zero. The sum rule for the rotational strengths also means that an average denominator approximation like that for $\alpha(\omega)$ in eq. (15.47) yields zero for the optical rotation.

15.7 Intermolecular forces

Only in gases at high temperatures and low pressures can the interactions between molecules be ignored. The non-ideality of gases, leading ultimately to liquefaction at low temperatures and high pressures, and the cohesion of liquids and solids show that molecules tend to attract one another at long distances. The relative incompressibility of liquids and solids shows that molecules tend to repel one another at shorter distances.

Much of this behaviour is understandable from classical electrostatics. In Section 8.5 we saw how permanent dipoles in a gas interact so as to lower their average energy. This occurs because the configurations of lowest energy are most favourably weighted by the Boltzmann factor. Making the molecules polarizable lowers the energy still further, since molecules will distort only if it is energetically favourable to do so. These energies become more negative as the molecules approach more closely, and so correspond to attractive forces. At shorter distances, the overlap between molecular charge distributions raises the energy more and more, so contributing to the repulsive forces.

However, certain features of the interaction between molecules can only be understood quantum-mechanically. For repulsions even this is difficult. Contributions arise from exchange forces when the charge clouds start to overlap; these can be viewed as a consequence of attempting to force electrons from one molecule into occupied orbitals on the other, in violation of the Pauli exclusion or antisymmetry principle. Such interactions are important at distances too large for the usual approximations of valence theory to be applicable. No entirely satisfactory quantitative theory of the repulsive interactions is yet available, and so we must confine the present section to attractive forces. Quantum mechanics shows how even neutral spherical atoms can attract one another. This attraction is clearly essential to account for the cohesion of the noble gas solids like neon and argon, but is inexplicable classically, at least at absolute zero. It is known as the *London* or *dispersion* force.

The dispersion force can be viewed as arising from the energy of interaction of instantaneous fluctuating dipole moments on the two atoms or molecules. How this arises can be seen from a simple model consisting of two harmonic oscillators of mass m and frequency ω_0 separated by a distance r. Each carries a charge $-q$ at the equilibrium position and an oscillating charge q. If the displacement operator of the first oscillator is $\hat{X}_1$, its dipole moment operator is $q\hat{X}_1$, and the electrostatic interaction between the two oscillators (assumed collinear for simplicity) is then described by the operator

$$\hat{V} = (q^2/2\pi\epsilon_0 r^3)\hat{X}_1\hat{X}_2. \tag{15.88}$$

Now the Hamiltonian for each oscillator is of the form

$$\hat{H}_j = \hat{P}_j^2/2m + \tfrac{1}{2}m\omega_0^2\hat{X}_j^2, \tag{15.89}$$

so that the total coupled Hamiltonian is $\hat{H}_1 + \hat{H}_2 + \hat{V}$. This can be separated into the sum of two independent harmonic oscillator Hamiltonians $\hat{H}_+ + \hat{H}_-$ depending on new momenta $\hat{P}_\pm = (\hat{P}_1 \pm \hat{P}_2)/2^{1/2}$ and coordinates $\hat{X}_\pm = (\hat{X}_1 \pm \hat{X}_2)/2^{1/2}$. Then

$$\hat{H}_\pm = \hat{P}_\pm^2/2m + \tfrac{1}{2}m\omega_0^2(1 \pm K)\hat{X}_\pm^2, \tag{15.90}$$

where

$$K = q^2/2\pi\epsilon_0 r^3 m\omega_0^2, \tag{15.91}$$

so that the new oscillator frequencies are

$$\omega_\pm = \omega_0(1 \pm K)^{1/2}. \tag{15.92}$$

Classically, the system has zero energy at zero temperature, but quantum-mechanically an oscillator must always have at least its zero-point energy. For the interacting system the zero-point energy is $\tfrac{1}{2}\hbar(\omega_+ + \omega_-)$, while for the non-interacting system it is $\tfrac{1}{2}\hbar\omega_0$ for each oscillator. The energy change due to the interaction is then

$$W = \tfrac{1}{2}\hbar(\omega_+ + \omega_- - 2\omega_0) \tag{15.93}$$

$$= \tfrac{1}{2}\hbar\omega_0[(1+K)^{1/2} + (1-K)^{1/2} - 2]. \tag{15.94}$$

At large enough distances r, the factor K will be small compared with unity and the square roots in eq. (15.94) can be expanded by the binomial theorem and truncated at terms of order K^2. The interaction energy then becomes

$$W = -\hbar\omega_0 K^2/8, \tag{15.95}$$

corresponding to an attractive interaction varying with distance as K^2 or $1/r^6$. Further terms in the expansion vary as higher even powers of K and are all attractive.

A more rigorous treatment of the dispersion force at large distances can be obtained using perturbation theory. The molecules interact through their charge distributions, each of which is expanded in a multipole series. One thus obtains a series of energy contributions in successive powers of the perturbation, which is itself the product of two series in inverse powers of the intermolecular distance r. Fortunately, in practice one is mainly concerned with the contributions of longest range, i.e. involving the lowest inverse powers of r. For neutral molecules, the leading term in the interaction operator is the dipole–dipole term

$$\hat{V} = [r^2 \boldsymbol{p}_1 \cdot \boldsymbol{p}_2 - 3(\boldsymbol{p}_1 \cdot \boldsymbol{r})(\boldsymbol{p}_2 \cdot \boldsymbol{r})]/4\pi\epsilon_0 r^5, \tag{15.96}$$

which varies as $1/r^3$. The dipole–quadrupole term varies as $1/r^4$ and the quadrupole–quadrupole term as $1/r^5$.

The zeroth-order states are simply superpositions of the states of the two non-interacting molecules. As only long-range interactions are to be considered, these states do not need to be antisymmetrized, and we write them as $|mn\rangle$, where m denotes that the first molecule is in state $|m\rangle$ and n that the second molecule is in state $|n\rangle$. The energy of the state $|mn\rangle$ is $E_{mn} = E_m + E_n$. We require the interaction between the two molecules in their ground states, i.e. when the system is in its ground state $|00\rangle$, with the unperturbed energy E_{00}.

The first-order perturbation energy is, from eq. (14.13),

$$W^{(1)} = \langle 00 | \hat{V} | 00 \rangle. \tag{15.97}$$

As noted in Chapter 14, this is simply the expectation value of the operator in the unperturbed ground state. With the leading interaction term $\hat{V}$ given by eq. (15.96), $W^{(1)}$ is then the ordinary electrostatic interaction between the permanent dipole moments (if any) of the two molecules. For a given orientation, this varies like $1/r^3$, but in a fluid the thermal average of the interaction over all orientations varies as the square of this, i.e. $1/r^6$ (see Section 8.5). Higher terms in $\hat{V}$ yield energies of interaction involving permanent quadrupole moments and other higher moments, and in the thermal average these vary as higher powers of $1/r$ than the sixth.

The second-order perturbation energy is, from eq.(14.26),

$$W^{(2)} = \sum_{mn(\neq 00)} \frac{|\langle 00 | \hat{V} | mn \rangle|^2}{E_{00} - E_{mn}}, \tag{15.98}$$

where one of the states $|m\rangle$ and $|n\rangle$ may be the molecular ground state, but not both. There are therefore two classes of terms in $W^{(2)}$: those involving singly-excited configurations, in which one of the molecules remains in the ground state, and those involving doubly-excited configurations, in which neither molecule remains in the ground state. The first class of term entails distortion of one molecule but not the other, while the second class of term involves distortions of both molecules together. As noted earlier, for the ground state the second-order perturbation energy must be stabilizing.

The terms involving single excitations of the first molecule are of the form

$$W_s^{(2)} = \sum_{m(\neq 0)} \frac{|\langle 00|\hat{V}|m0\rangle|^2}{E_0 - E_m}, \tag{15.99}$$

since there is no contribution to the excitation energy from the second molecule. For the leading dipole–dipole term in the interaction, $\hat{V}$ is given by eq. (15.96) and the matrix element in eq. (15.99) becomes

$$\langle 00|\hat{V}|m0\rangle = [r^2 \boldsymbol{p}_{0m}\cdot\boldsymbol{p}_0 - 3(\boldsymbol{p}_{0m}\cdot\boldsymbol{r})(\boldsymbol{p}_0\cdot\boldsymbol{r})]/4\pi\epsilon_0 r^5, \tag{15.100}$$

where the transition dipole moment matrix element $\boldsymbol{p}_{0m}$ refers to the first molecule and the permanent dipole moment $\boldsymbol{p}_0$ to the second molecule. When this result is substituted in eq. (15.99), each of the terms for a given m involves a product of two transition dipole moment matrix elements divided by minus the corresponding excitation energy. The sum over m then yields the static polarizability of the first molecule (cf. eq. (15.23)). It is then found that eq. (15.99) gives the induction energy, i.e. the energy of interaction between the permanent dipole moment on the second molecule and the dipole moment this induces on the second molecule. The terms involving single excitations of the second molecule give similarly the energy associated with the dipole moment induced on the second molecule by the permanent dipole moment of the first molecule. These terms are already included in the electrostatic treatment of Section 8.5, and again vary as $1/r^6$.

The dispersion forces must therefore arise from the terms involving double excitations, as implied in the earlier remarks about instantaneous fluctuating dipole moments. The dispersion energy is

$$W_d = \sum_{\substack{m(\neq 0)\\ n(\neq 0)}} \frac{|\langle 00|\hat{V}|mn\rangle|^2}{E_{00} - E_{mn}}, \tag{15.101}$$

where the matrix element is

$$\langle 00|\hat{V}|mn\rangle = \{r^2 \boldsymbol{p}_{0m}^{(1)}\cdot\boldsymbol{p}_{0n}^{(2)} - 3[\boldsymbol{p}_{0m}^{(1)}\cdot\boldsymbol{r}][\boldsymbol{p}_{0n}^{(2)}\cdot\boldsymbol{r}]\}/4\pi\epsilon_0 r^5, \tag{15.102}$$

involving transition dipole moment matrix elements for both molecules. In a fluid, the energy W_d has to be averaged over all possible orientations of both molecules. This depends on averages for each molecule of the form

$$\langle(\boldsymbol{p}_{0m})_\alpha(\boldsymbol{p}_{m0})_\beta\rangle = \tfrac{1}{3}|\boldsymbol{p}_{0m}|^2\,\delta_{\alpha\beta}, \tag{15.103}$$

where α and β are Cartesian components. These averages are zero when $\alpha \neq \beta$ because positive and negative values are equally probable, and when $\alpha = \beta$ are the same for all three values of α and hence are equal to one-third of the sum of these three terms (as in the passage from eq. (15.22) to eq. (15.25)). Substitution of eq. (15.102) in W_d using eq. (15.103) leads with careful book-keeping to the result

$$W_d = -\frac{1}{24\pi^2\epsilon_0^2 r^6} \sum_{\substack{m(\neq 0)\\ n(\neq 0)}} \frac{|p_{0m}^{(1)}|^2 \, |p_{0n}^{(2)}|^2}{E_{mn} - E_{00}}. \tag{15.104}$$

Being basically a dipole–dipole term, this has the familiar $1/r^6$ dependence.

The appearance of the restricted sums, transition dipole moment matrix elements and energy denominator in eq. (15.104) suggests that W_d might be related in some way to the molecular polarizability, which also contains these factors. A simple way of doing this is to note that the energy denominator is the sum of the molecular excitation energies ΔE_m and ΔE_n and to invoke the average denominator approximation to write these as Δ_1 and Δ_2 to obtain

$$W_d \approx -\sum_{\substack{m(\neq 0)\\ n(\neq 0)}} |p_{0m}^{(1)}|^2 \, |p_{0n}^{(2)}|^2 / 24\pi^2\epsilon_0^2 r^6(\Delta_1 + \Delta_2). \tag{15.105}$$

In the same approximation the static polarizability from eq.(15.25) is

$$\alpha_j = 2 \sum_{m(\neq 0)} |p_{0m}^{(j)}|^2 / 3\Delta_j, \tag{15.106}$$

so that W_d can be expressed as

$$W_d \approx -3\alpha_1\alpha_2\Delta_1\Delta_2/(\Delta_1 + \Delta_2)32\pi^2\epsilon_0^2 r^6. \tag{15.107}$$

If the average excitation energies are approximated as the molecular ionization energies I_1 and I_2, eq. (15.107) becomes London's original expression for the dispersion energy. From this expression we see that more polarizable molecules are attracted to one another more strongly than molecules with tightly-bound less polarizable electron clouds. The more polarizable molecules are those with low-energy excitations of high oscillator strength. The average energy denominator approximation also implies that α is proportional to the number of electrons Z via the oscillator strength sum rule (eqs (15.45) and (15.47)), so that the larger molecules in a homologous series tend to attract one another more strongly. This and their higher mass lead to decreased volatility.

Another way of expressing W_d in terms of the molecular polarizabilities, this time exactly, makes use of the identity

$$\frac{1}{a+b} = \frac{2}{\pi} \int_0^\infty \frac{ab \, dx}{(a^2 + x^2)(b^2 + x^2)}. \tag{15.108}$$

We set a equal to the transition frequency $\omega_{m0} = \Delta E_m/\hbar$ and b equal to

$\omega_{n0} = \Delta E_n/\hbar$. Then W_d can be written as

$$W_d = -\frac{1}{12\pi^3\epsilon_0^2 r^6\hbar}\int_0^\infty dx \left\{\sum_{m(\neq 0)} \frac{\omega_{m0}\,|p_{m0}^{(1)}|^2}{\omega_{m0}^2 + x^2} \right.$$

$$\left. \times \sum_{n(\neq 0)} \frac{\omega_{n0}\,|p_{n0}^{(2)}|^2}{\omega_{n0}^2 + x^2}\right\}. \tag{15.109}$$

Each of the restricted sums can be replaced using eq. (15.25) by a polarizability at the *imaginary* frequency ix. This gives the dispersion energy as

$$W_d = -(3\hbar/16\pi^3\epsilon_0^2 r^6)\int_0^\infty \alpha^{(1)}(ix)\alpha^{(2)}(ix)\,dx. \tag{15.110}$$

This expression has the advantage of separating the energy into contributions from the two molecules. The polarizability at imaginary frequency is also a well-behaved function of x; unlike the contributions to the polarizability at real frequency with their rapid variation when the frequency is close to a transition frequency (cf. Fig. 9.1), all the contributions to $\alpha(ix)$ decrease monotonically as x goes from 0 to ∞, as sketched in Fig. 15.2. Given a set of theoretical or experimental transition dipole moment matrix elements and transition frequencies for the polarizability at real frequency, one can use eq.

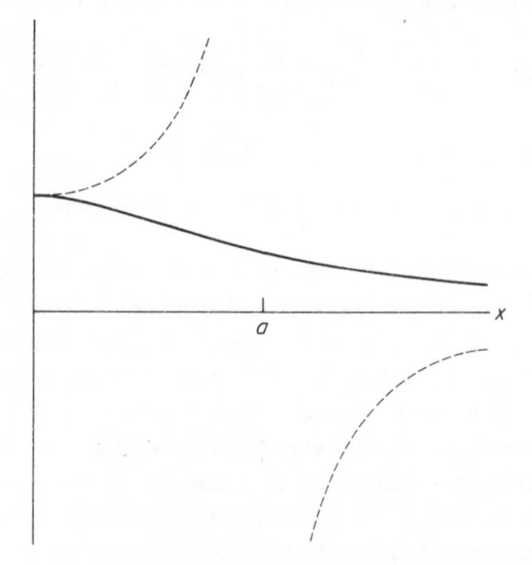

Fig. 15.2 The variation with x of the functions $1/(a^2 + x^2)$ (solid line) and $1/(a^2 - x^2)$(broken line). The latter function corresponds to the behaviour of terms in the usual polarizability $\alpha(x)$, ignoring damping, and the former function to the behaviour of terms in $\alpha(ix)$

(15.110) to calculate the dispersion interaction without further information. This link between the spectra and the energy of an interacting system is an important theoretical result, useful, for instance, in helping to understand the cohesion of molecular crystals.

Intermolecular forces and energies are not usually measured directly. Observable quantities are normally complicated thermal averages over separation and orientation. They include contributions not only from the leading term in the dispersion energy, conventionally written as $-C_6/r^6$, but also from higher terms such as $-C_8/r^8$ arising from dipole–quadrupole fluctuations and $-C_{10}/r^{10}$ arising from dipole–octupole and quadrupole–quadrupole fluctuations. Perhaps the most direct measurements have been made using delicate balances to determine the forces between samples of mica.

Micas are sheet silicates with a characteristic layer structure leading to easy and essentially perfect cleavage. They have the particular advantage of being smooth on a molecular scale, so perfect are the layers. The interaction between macroscopic samples can then be expressed in terms of their geometrical arrangement without having to average over an unknown distribution of distances. The total energy of attraction between macroscopic particles interacting with an energy $-C_6/r^6$ between their constituent microscopic particles has been evaluated for various geometries. For two semi-infinite plates the total energy varies as $1/d^2$, where d is the separation. For two spheres with their surface separated by d, with d much less than their radius, the energy varies as $1/d$, and the same dependence is found for a sphere and a plane or two crossed cylinders. It is noteworthy that the very short-ranged interaction between individual elements can nevertheless lead to much longer-ranged interaction between macroscopic particles. This helps to account for flocculation in colloidal systems, and the dependence of the interaction on geometry may influence self-assembly in biological systems.

The geometries used for the measurements on micas include a sheet bent into a half-cylinder interacting with a plane sheet, and two half-cylinders with their axes perpendicular. The results of these measurements are consistent with the expected $1/r^6$ dependence of the leading contribution to the energy at short distances. However, at longer distances of the order of 10 nm the results are consistent with a change to a leading $1/r^7$ dependence expected from a more careful analysis of long-range effects.

The origin of the change in distance dependence lies in the finite time taken for electromagnetic interactions to propagate, their speed being limited by the speed of light *in vacuo*, c. Modifications to allow for this must be made much as in the retarded potentials discussed in Section 6.10. The fluctuation in the first molecule at time t is experienced at the second molecule at the later time $[t] = t + r/c$. This induces a retarded dipole moment on the second molecule which interacts with the original fluctuation at time $t + 2r/c$. If the time lag $2r/c$ is negligible compared with the time-scale for the fluctuations, which is essentially the inverse transition frequency or roughly $\hbar/\Delta$, the interaction can be regarded as instantaneous. Electrostatic calculations as performed earlier in this section thus suffice at small enough distances r. However, at longer distances the original fluctuation has decayed significantly by the time the dipole moment it induced has reacted back. The interaction is thereby weakened, and the distance dependence changes to $1/r^7$. The observation of this change provides excellent experimental proof of retardation, the critical distance being compatible with excitation energies of a few eV as expected. Similar effects occur in the interaction of other large systems such as colloidal particles.

Examples

15.1 A molecule has only one important electronic transition, of oscillator strength 1.00 at a wavelength of 200 nm. Calculate the static molecular polarizability.

15.2 The molecule in Example 15.1 crystallizes with one molecule per unit cell of side 300 pm. Use the results of Section 9.3 to calculate the static dielectric constant of the crystal, and its refractive index for light of wavelength 400 nm and 600 nm.

15.3 A harmonic oscillator of mass m, frequency ω_0 and charge q perturbed by an electric field E is described by the Hamiltonian

$$\hat{H} = -(\hbar^2/2m)(\mathrm{d}^2/\mathrm{d}x^2) + \tfrac{1}{2}m\omega_0^2x^2 - qEx.$$

In zero field the two lowest eigenstates are

$$|0\rangle = \pi^{-1/4}\exp(-\tfrac{1}{2}\xi^2)$$
$$|1\rangle = 2^{1/2}\pi^{-1/4}\xi\exp(-\tfrac{1}{2}\xi^2),$$

where $\xi = (m\omega_0/\hbar)^{1/2}x$. Verify that the electric field perturbation does not change the energy of these states to first order. Calculate the second-order change in the ground-state energy, given that $\langle 0|\xi|n\rangle$ is zero for $n > 1$ and that

$$\int_0^\infty \exp(-\xi^2)\,\mathrm{d}\xi = \tfrac{1}{2}\pi^{1/2}.$$

Hence calculate the static polarizability of the harmonic oscillator.

15.4 By changing the variable in $\hat{H}$ in Example 15.3 to

$$y = x - qE/m\omega_0^2$$

show that $\hat{H}$ describes a harmonic oscillator of frequency ω_0, but with shifted energy levels. Hence calculate the polarizability and compare it with that obtained in Example 15.3.

15.5 A hydrogen atom in the ground state $|0\rangle$ is subjected to the electric field perturbation $\hat{V} = -eEz$. The first-order perturbation energy $W_0^{(1)}$ is zero, and the second-order perturbaton energy $W_0^{(2)}$ is given in terms of the first-order change in the state $|v_0^{(1)}\rangle$ by eq. (14.24) as $W_0^{(2)} = \langle 0|\hat{V}|v_0^{(1)}\rangle$. Given that the normalized wavefunction and its first order change are

$$|0\rangle = (\pi a^3)^{-1/2}\exp(-r/a)$$
$$|v_0^{(1)}\rangle = (eEma/2\hbar^2)z(r+2a)|0\rangle$$

where a is the Bohr radius $4\pi\epsilon_0\hbar^2/me^2$, calculate $W_0^{(2)}$ and hence show that the polarizability is $18\pi\epsilon_0 a^3$. The integrals required are of the form

$$\int_0^\infty r^n \exp(-2r/a)\,\mathrm{d}r = (\tfrac{1}{2}a)^{n+1}\,n!.$$

15.6 Calculate the numerical value of the polarizability in the ground state of the hydrogen atom in SI units, esu, and atomic units (where the unit of length is the Bohr radius).

15.7 The energy $W_0^{(2)}$ in Example 15.5 can also be calculated by standard perturbation theory. By symmetry, the only non-zero matrix elements of $\hat{V}$ link the 1*s* state $|0\rangle$ with the np_z states. Calculate the contribution to $W_0^{(2)}$ from the $2p_z$ state, which has the wavefunction

$$|2p_z\rangle = (32\pi a^5)^{-1/2} z \exp(-r/2a)$$

and excitation energy $-3\hbar^2/8ma^2$. Given that this contribution is 82 per cent of the total from all np_z states, calculate the total. Why does it differ from the exact value obtained in Example 15.5?

15.8 Calculate the diamagnetic magnetizability $\varkappa_d$ for the hydrogen atom in the ground state $|0\rangle$. The required wavefunction and intergrals are given in Example 15.5.

Further reading

P. W. Atkins, *Molecular Quantum Mechanics,* 2nd edn, Clarendon Press, Oxford, 1983.

J. H. Van Vleck, *The Theory of Electric and Magnetic Susceptibilities*, Oxford University Press, *1965*.

D. W. Davies, *The Theory of the Electric and Magnetic Properties of Molecules*, Wiley, London, 1967.

J. L. Friar and S. Fallieros, Diamagnetism, gauge transformations, and sum rules, *Amer. J. Phys.*, **49** (1981) 847.

J. Mahanty and B. W. Ninham, *Dispersion Forces,* Academic Press, London, 1976.

Chapter 16

Theory of spectroscopy

Several matters relevant to spectroscopy have already been considered. In Chapter 12 there was a phenomenological treatment of aspects of the quantum nature of radiation, which reveals itself when absorption and emission take place. In Chapter 13 we obtained the Hamiltonian for a charged particle in an electromagnetic field, this being the origin of the interaction causing stimulated absorption and emission. Then in Chapter 14 we examined the basic principles governing the change of quantum state under a periodic perturbation such as that provided by an electromagnetic wave.

We now show how the Einstein coefficients are related to fundamental molecular properties. This permits certain general statements about when the coefficients can be non-zero, i.e. selection rules. We are, however, not concerned here with the details of spectroscopic measurement and interpretation but with the electromagnetic and quantum-mechanical principles governing the processes which spectroscopy observes. Such principles are often glossed over in specialized texts on spectroscopy, which cannot rely on the background material in electromagnetism presented here. As well as treating the principles of ordinary absorption and emission, we also give a brief account of light scattering.

16.1 Interaction Hamiltonian

The Hamiltonian of eq. (13.35) gives the interaction of a charge with an electromagnetic field in terms of the vector and scalar potentials. We first express the potentials in terms of other quantities characterizing an electromagnetic wave. Then later the rate of absorption will turn out to be expressible in terms of the photon density, as required in order to obtain the Einstein coefficient. In the Coulomb gauge, an electromagnetic wave of radian frequency ω propagating in the direction $\boldsymbol{n}$ is fully described by a vector potential

$$\boldsymbol{A} = \boldsymbol{A}_0 \cos\,[\omega(t - \boldsymbol{n}\cdot\boldsymbol{r}/c)] \tag{16.1}$$

at a point $\boldsymbol{r}$ and time t. The external potential V_e can be taken as a constant which has no effect on the transitions and so can be absorbed in the definition of the zero of energy. This result can also be obtained from eq. (13.33) by applying the Lorentz gauge, which gives wave equations for both potentials in free space, and then using the residual freedom in the choice of gauge to make the scalar potential zero. We shall treat only low-intensity light waves such that no non-linear effects occur (though these are of course important in some applications, e.g. lasers). The term in A^2 can then be neglected, leaving

$$\hat{H}_{\mathrm{f}} = (i\hbar q/m)\boldsymbol{A}\cdot\nabla = -(q/m)\boldsymbol{A}\cdot\hat{\boldsymbol{P}}, \tag{16.2}$$

where $\hat{\boldsymbol{P}}$ is the momentum operator as usual.

The amplitude vector $\boldsymbol{A}_0$ for the vector potential can be related to the energy density of the wave, i.e. to its intensity. This is conveniently achieved via the electric field. Recall that in an electric field $\boldsymbol{E}$ the energy density is $\frac{1}{2}\epsilon_0 E^2$, which for an electromagnetic wave equals the energy density in the magnetic induction. The total is therefore $\epsilon_0 E^2$, which has to be averaged over one cycle of the oscillation. For a uniform scalar potential

$$\boldsymbol{E} = -\partial\boldsymbol{A}/\partial t \tag{16.3}$$

$$= \boldsymbol{A}_0\omega \sin\,[\omega(t - \boldsymbol{n}\cdot\boldsymbol{r}/c)]. \tag{16.4}$$

Since the average of $\sin^2\theta$ over one cycle is $\frac{1}{2}$, the energy density is then

$$w = \tfrac{1}{2}\epsilon_0\omega^2 A_0^2 \tag{16.5}$$

so that

$$A_0 = (2w/\epsilon_0\omega^2)^{1/2}. \tag{16.6}$$

Here w can be expressed in terms of the photon thermal population $N(\omega, T)$ as

$$w = N(\omega, T)\hbar\omega/\Omega \tag{16.7}$$

where Ω is the volume.

We can therefore write the interaction Hamiltonian in the form

$$\hat{H}_{\mathrm{f}}(t) = -(q/m)(2w/\epsilon_0\omega^2)^{1/2} \cos\,[\omega(t - \boldsymbol{n}\cdot\hat{\boldsymbol{R}}/c)]\,\boldsymbol{e}\cdot\hat{\boldsymbol{P}}. \tag{16.8}$$

Here $\hat{\boldsymbol{R}}$ is the usual position operator and $\boldsymbol{e}$ is a polarization unit vector parallel to $\boldsymbol{A}$ and $\boldsymbol{E}$. Equation (16.8) shows that $\hat{H}_{\mathrm{f}}(t)$ takes the form

$$\hat{H}_{\mathrm{f}}(t) = \hat{V}\mathrm{e}^{-i\omega t} + \hat{V}^{\dagger}\mathrm{e}^{i\omega t}, \tag{16.9}$$

with the dagger denoting the Hermitian conjugate or adjoint operator. This form is a slight generalization of that used to discuss transitions under the influence of a periodic perturbation in Section 14.5.

16.2 Transition rate

We shall calculate the rate of absorption, which we know is governed by the same Einstein coefficient as the stimulated emission. The absorption is determined by the term in $\mathrm{e}^{-i\omega t}$, and for a density of photon states $\varrho(\omega)$ proceeds at a rate

$$W = (2\pi/\hbar^2)\,|V_{\mathrm{ul}}|^2\,\varrho(\omega). \tag{16.10}$$

Here u and l denote the upper and lower states between which absorption occurs for radiation of frequency ω, and the required matrix element is

$$V_{\mathrm{ul}} = -\left(\frac{q}{2m}\right)\left(\frac{2w}{\epsilon_0\omega^2}\right)^{1/2}\langle u\,|\,\mathrm{e}^{-i\omega\boldsymbol{n}\cdot\hat{\boldsymbol{R}}/c}\boldsymbol{e}\cdot\hat{\boldsymbol{P}}\,|\,l\rangle. \tag{16.11}$$

(Compared with eq. (14.83), eq. (16.10) has an extra factor of $\hbar$ in the denominator because it uses a density of states per unit frequency interval instead of per unit energy interval.)

Now molecular dimensions range from fractions of a nm upwards, so that a typical order of magnitude for molecular size would be 1 nm (excluding macromolecules). This is very small compared with optical wavelengths of several hundred nm, and *a fortiori* compared with the longer wavelengths of infrared and microwave radiation. It follows that the exponential in eq. (16.11) varies little over the region where the molecular wavefunction differs from zero. We can therefore expand the exponential about a suitable origin in the molecule and retain only the leading term, unity. The matrix element then starts with the term

$$V_{\mathrm{ul}}^{(0)} = -\frac{q}{m\omega}\left(\frac{w}{2\epsilon_0}\right)^{1/2}\langle u\,|\,\boldsymbol{e}\cdot\hat{\boldsymbol{P}}\,|\,l\rangle. \tag{16.12}$$

The (electric) *dipole approximation* consists of approximating the whole matrix element by this one term. It is so called because $V_{\mathrm{ul}}^{(0)}$ is found to be proportional to the corresponding matrix element of the electric dipole operator $q\hat{\boldsymbol{R}}$, as shown below. Higher terms in the expansion of the exponential give other electric and magnetic multipole terms, as outlined below.

Equation (16.12) is expressed in terms of a transition dipole moment matrix element by means of the result (15.38), which in the present case takes the form

$$\langle u\,|\,\hat{\boldsymbol{P}}\,|\,l\rangle = im\omega\langle u\,|\,\hat{\boldsymbol{R}}\,|\,l\rangle. \tag{16.13}$$

When this result is substituted in eq. (16.12), the factor $m\omega$ cancels and the transition dipole moment matrix element $\boldsymbol{p}_{\mathrm{ul}}$ defined in eq. (15.17) is introduced. The matrix element then becomes

$$V_{\mathrm{ul}}^{(0)} = -i(w/2\epsilon_0)^{1/2}\boldsymbol{e}\cdot\boldsymbol{p}_{\mathrm{ul}}, \tag{16.14}$$

so that for electric dipole transitions the rate (16.10) becomes

$$W = (\pi w/\epsilon_0\hbar^2)\,|\,\boldsymbol{e}\cdot\boldsymbol{p}_{\mathrm{ul}}\,|^2\,\varrho(\omega). \tag{16.15}$$

If the transition dipole moment matrix element is zero, a non-zero transition rate may still arise from the next term in the expansion of the exponential in eq. (16.11). This yields a matrix element $V_{\mathrm{ul}}^{(1)}$ proportional to

$$\langle u\,|\,i(\omega/c)(\boldsymbol{n}\cdot\hat{\boldsymbol{R}})(\boldsymbol{e}\cdot\hat{\boldsymbol{P}})\,|\,l\rangle. \tag{16.16}$$

Suppose for definiteness that the light wave is propagating in the z direction and polarized in the x direction so that $\boldsymbol{n}\,||\,z$ and $\boldsymbol{e}\,||\,x$. The matrix element is then $(i\omega/c)\langle u|\hat{Z}\hat{P}_x|l\rangle$, where we can write

$$\hat{Z}\hat{P}_x = \tfrac{1}{2}(\hat{Z}\hat{P}_x + \hat{X}\hat{P}_z) + \tfrac{1}{2}(\hat{Z}\hat{P}_x - \hat{X}\hat{P}_z). \tag{16.17}$$

The last term here is half the angular momentum operator $\hat{L}_y$, giving

$$V_{\rm ul}^{(1)} \sim (i\omega/2c)[\langle u|\hat{Z}\hat{P}_x + \hat{X}\hat{P}_z|l\rangle + \langle u|\hat{L}_y|l\rangle], \tag{16.18}$$

where it may be shown by arguments like those which lead to the relation (16.13) that

$$\langle u|\hat{Z}\hat{P}_x + \hat{X}\hat{P}_z|l\rangle = -im\omega\langle u|\hat{Z}\hat{X}|l\rangle. \tag{16.19}$$

Thus $V_{\rm ul}^{(1)}$ can be written as the sum of two terms, one depending on *electric quadrupole* matrix elements like $\langle u|q\hat{Z}\hat{X}|l\rangle$ and the other depending on *magnetic dipole* matrix elements like $\langle u|(q/2m)\hat{L}_y|l\rangle$. If the electric dipole moment matrix element is non-zero, its contribution $V_{\rm ul}^{(0)}$ will exceed $V_{\rm ul}^{(1)}$ from these other terms by an amount of the order of the ratio between the wavelength of the light and the size of the molecule, from eq. (16.16). For optical and longer wavelengths, this ratio is typically three or more orders of magnitude.

16.3 Einstein coefficients

The transition rate for electrical dipole transitions contains the combination $w\varrho(\omega)$, where $\varrho(\omega)$ is the density of photon states per unit radian frequency interval. If we change to the circular frequency $\nu = \omega/2\pi$, the corresponding density of photon states is $\varrho(\nu) = 2\pi\varrho(\omega)$, which is just the volume Ω times the density of states per unit frequency per unit volume $n(\nu)$ from eq. (12.14). We can therefore write

$$\varrho(\omega) = \Omega n(\nu)/2\pi \tag{16.20}$$

and using eq. (16.7) for w we obtain

$$w\varrho(\omega) = h\nu N(\nu, T)n(\nu)/2\pi \tag{16.21}$$

$$= \varrho(\nu, T)/2\pi, \tag{16.22}$$

where $\varrho(\nu, T)$ is the black-body distribution of eq. (12.16).

The total rate of transitions from the lower state of a molecule to the upper is, from eq. (16.15),

$$W = (1/2\epsilon_0\hbar^2)|\boldsymbol{e}\cdot\boldsymbol{p}_{\rm ul}|^2\,\varrho(\nu, T). \tag{16.23}$$

If there are $N_{\rm l}$ molecules in the lower state, each will make transitions at the rate W, and then the rate of decrease of $N_{\rm l}$ becomes

$$-{\rm d}N_{\rm l}/{\rm d}t = N_{\rm l}W. \tag{16.24}$$

Comparison with eq. (12.19) now shows that the Einstein coefficient for stimulated absorption is given by

$$B_{\rm lu}(\nu) = |\boldsymbol{e}\cdot\boldsymbol{p}_{\rm ul}|^2/2\epsilon_0\hbar^2. \tag{16.25}$$

The final form of the Einstein coefficient describing a particular experiment depends on the polarization of the radiation and on the orientation of the transition dipole moment, as eq. (16.25) shows. Thus, for example, spectra of molecular crystals taken using plane-polarized light show different intensities for different relative orientations of the crystal axes and the plane of polarization, the details depending on the arrangement of the molecules in the crystal structure. Frequently, however, spectra are taken in fluids where the molecules may be regarded as randomly reorienting. It is then necessary to average the factor $|\boldsymbol{e}\cdot\boldsymbol{p}_{\mathrm{ul}}|^2$. If the transition dipole moment makes an angle θ with the unit vector $\boldsymbol{e}$ describing the polarization of the radiation, the required average is $p_{\mathrm{ul}}{}^2\,\langle\cos^2\theta\rangle$. The trigonometrical average can be evaluated using the weighting factor $\sin\theta$ as in Fig. 9.2. Alternatively one notes that $\langle\cos^2\theta\rangle = \langle z^2/r^2\rangle$ which in a random average must also equal $\langle x^2/r^2\rangle$ and $\langle y^2/r^2\rangle$; since $x^2+y^2+z^2=r^2$, the sum of these three averages must equal unity, so that each equals 1/3. If the radiation is not pure plane polarized, each component gives the same result. The mean Einstein coefficient then becomes

$$\bar{B}_{\mathrm{lu}}(\nu) = p_{\mathrm{ul}}{}^2/6\epsilon_0\hbar^2. \tag{16.26}$$

The corresponding coefficient for stimulated emission is the same, as we saw in Chapter 12, and the coefficient for spontaneous emission follows from eq. (12.26) as

$$\bar{A}_{\mathrm{ul}}(\nu) = 8\pi^2\nu^3 p_{\mathrm{ul}}^2/3\epsilon_0\hbar c^3, \tag{16.27}$$

Since the Einstein coefficients depend on the square of the transition dipole moment matrix element, like the polarizability they can be rewritten in terms of the oscillator strength f defined in eq. (15.26). For stimulated absorption the result is

$$\bar{B}_{\mathrm{lu}}(\nu) = e^2 f_{\mathrm{ul}}/12\epsilon_0 m_e h\nu \tag{16.28}$$

and for spontaneous emission it is

$$\bar{A}_{\mathrm{ul}}(\nu) = 2\pi\nu^2 e^2 f_{\mathrm{ul}}/3\epsilon_0 m_e c^3. \tag{16.29}$$

As $\bar{B}_{\mathrm{lu}}(\nu)$ determines the absorption coefficient, this too can be expressed in terms of the oscillator strength. For the absorption intensity integrated with respect to wavenumber, the result is

$$A' = \bar{B}_{\mathrm{lu}} L h\nu/c^2 \tag{16.30}$$

$$= Le^2 f_{\mathrm{ul}}/12\epsilon_0 m_e c^2, \tag{16.31}$$

where the coefficient of f_{ul} is equal to $1.77\times10^9\ \mathrm{m\,mol^{-1}}$. The second expression has the advantage that the oscillator strength completely determines the integrated intensity whereas the Einstein coefficient does so only in conjunction with the frequency $\dot{\nu}$. We therefore see that transitions of high oscillator strength are highly absorbing at the frequency in question, making the molecule highly coloured if this frequency lies in the visible region of the spectrum. Such transitions are also major contributors to the molecular polarizability and through that to the intermolecular forces. The tendency of dye molecules to form aggregates can thus be understood.

16.4 Selection rules and intensity

The rate of absorption in an electric dipole transition depends on the square of the transition dipole moment matrix element between the upper and lower states. The states may be of all kinds, electronic, vibrational, rotational, vibronic, rovibronic, and so on, but in each case the intensity depends on differences in electronic and nuclear charge distributions in the two states. In particular, the overall appearance of the spectrum is determined by those transitions with non-zero matrix elements p_{ul}. The conditions which fix when p_{ul} can be non-zero are known as *selection rules*. These are of two sorts: *gross* selection rules which determine whether there is any activity at all in the type of spectrum in question, and *specific* selection rules which determine which pairs of states are active. A general discussion of selection rules requires a detailed consideration of molecular states and use of group theory, both of which lie outside the scope of this book, and we therefore concentrate on a few basic principles.

One particular point to notice is that the photon has an angular momentum of unity, so that the angular momentum of a molecule must change by one unit on absorption or emission of a photon, as angular momentum is conserved. If the energy eigenstates are also eigenstates of angular momentum, specific selection rules emerge at once. A simple example is provided by the hydrogen atom, where electronic transitions are subject to the selection rule $\Delta l = \pm 1$ and $\Delta m = 0, \pm 1$, where l is the quantum number for the orbital angular momentum and m is that for its z component. Thus $s \rightarrow s$ transitions are forbidden, as are $p \rightarrow p$, $d \rightarrow d$, and so on. As states differing only in the value of m are degenerate, the selection rule on m has no effect on the energy of the allowed transitions. It does, however, govern the polarization of the photon which is absorbed or emitted: $\Delta m = 0$ gives linear polarization and $\Delta m = \pm 1$ gives right and left circular polarizations.

An important selection rule governs all electric dipole transitions in molecules and crystals possessing a centre of symmetry. In such systems, all states are either even or *gerade*, remaining unchanged under inversion, or else odd or *ungerade*, changing sign under inversion. The dipole operator is always odd, since it changes sign under inversion. Then if the upper and lower states are both even or both odd (i.e. of the same *parity*), the whole matrix element p_{ul} changes sign under inversion. Since it must also remain unchanged under a symmetry operation like inversion, it can only be zero. Thus in centrosymmetric systems electric dipole transitions are allowed only between states of opposite parity. This is the *Laporte* selection rule. For the hydrogen atom it gives a result consistent with that deduced above on different grounds, since orbitals with even values of l transform like even powers of the coordinates and hence are even under inversion, and orbitals with l odd are odd.

Selection rules need not be completely rigorous. The classification of a transition as forbidden may rest on an approximate description of the states concerned, for instance one in which certain weak couplings are neglected. Once these couplings are included, the transition may become weakly allowed, i.e. of low intensity. For example, if spin-orbit coupling is neglected electronic states can be classified by their spin multiplicity. Since the dipole moment operator does not affect the spin state, allowed electronic transitions must entail no change of spin. Such transitions are normally intense: the colours of dyes arise from transitions with oscillator strengths of 1 or more, while the purple colour of the manganate(VII) or permanganate ion MnO_4^- has $f = 0.03$. Non-zero spin-orbit coupling allows spin-forbidden transitions to occur, with oscillator strengths of typically 10^{-4} for *d–d* transitions in transition metal ions and 10^{-6} or less for *f–f* transitions in rare earth ions.

For a strong electronic transition, the transition dipole moment might be equivalent to moving one electronic charge through 0.1 nm (1 Å), so that $p_{ul} = 1.6 \times 10^{-29}$ C m. Then the mean Einstein coefficient $\bar{B}_{ul}$ is found to equal $6.0 \times 10^{36}\ m^3\ J^{-1}\ s^{-2}$. At a wavelength of 550 nm, in the middle of the visible spectrum, the integrated intensity is found to be $1.0 \times 10^9\ m\ mol^{-1}$, corresponding to an oscillator strength of 0.56. If this intensity is distributed over a band of width $\Delta\nu$, the mean molar absorption coefficient is given by eq. (12.41) as $A'c/\Delta\nu \ln 10$. For a wavenumber width $\Delta\nu$ of $1000\ cm^{-1}$, this expression yields $4300\ m^2\ mol^{-1}$.

Vibrational transitions entail changing the energy in a given normal mode (ignoring anharmonic coupling between modes). This changes the amplitude of vibration. The dipole moment amplitude in the two levels will therefore change and lead to a non-zero transition dipole moment matrix element, provided the mode in question entails a variation of dipole moment: this is the gross selection rule. So for example all heteronuclear diatomic molecules and no homonuclear diatomic molecules exhibit pure vibrational absorption. The specific selection rule for harmonic vibrations is found to be $\Delta v = \pm 1$, where v is the vibrational quantum number. This is plausibly consistent with the angular momentum selection rule, since the projection of a point moving uniformly around the circumference of a circle with fixed angular momentum performs simple harmonic motion along a diameter. The rule is also compatible with the Laporte selection rule, since harmonic vibrational states have the same parity as the quantum number v.

Vibrational transitions usually entail smaller transition dipole moments than allowed electronic transitions, since vibrational amplitudes are of the order of pm. This reduces the integrated intensity and oscillator strength by three orders of magnitude or more compared with electronic transitions. However, the bandwidth may be reduced similarly to $1\ cm^{-1}$ or less, so that the mean molar absorption coefficient may still be of the order of $1000\ m^2\ mol^{-1}$.

For rotational transitions, the selection rules can be simply rationalized much as for vibrational transitions. A rotational transition to a higher state accelerates the rotational motion and hence the angular variation of the dipole moment, provided there is a non-zero permanent electric dipole moment: this is the gross selection rule. The specific selection rule for rigid rotors is found to be $\Delta J = \pm 1$, where J is the total angular momentum quantum number. For rigid linear molecules this is the entire selection rule. For symmetric tops (molecules with two principal moments of inertia equal) there is an additional selection rule $\Delta K = 0$, where K is the quantum number describing the angular momentum about the symmetry axis. The permanent dipole moment necessarily lies along this axis, so that rotation about the axis entails no variation in dipole moment and cannot interact with an electromagnetic wave. Asymmetric tops (with three different principal moments of inertia) have no additional selection rules, while spherical tops (with three principal moments of inertia equal) necessarily have no permanent dipole moment.

The gross selection rules forbid absorption or emission between pure non-polar vibrational states and for non-polar molecules between pure rotational states. However, an allowed electronic transition may be accompanied by otherwise forbidden vibrational and rotational transitions, and an allowed vibrational transition may be accompanied by otherwise forbidden rotational transitions: the transition dipole moment matrix element has to be non-zero between the total upper and lower states and not between each component of these states individually. Thus the vibrational and rotational states of homonuclear diatomic molecules can be studied through the fine structure of their electronic spectra. Similarly the rotational states of a centrosymmetric linear molecule like carbon dioxide can be studied through the fine structure of vibrational spectra involving the bending mode or the asymmetric stretching mode.

In Section 16.2 electric quadrupole and magnetic dipole transitions were mentioned. Because these depend on different matrix elements, they are subject to different selection rules. The electric quadrupole operator is even under inversion, and therefore has non-zero matrix elements between states of the same parity. For systems in a spherically symmetric potential, such as the hydrogen atom, the specific selection rule for electric quadrupole transitions is $\Delta l = 0, \pm 2$ (except that transitions between states both having $l = 0$ are forbidden) and $\Delta m = 0, \pm 1, \pm 2$.

In magnetic dipole transitions in systems with zero spin ($S = 0$), the angular momentum operators can only change the z component of angular momentum. In a spherically symmetric potential, this is an eigenvalue of $\hat{L}_z$, but $\hat{L}_x$ and $\hat{L}_y$ can change the quantum number m by ± 1. These sublevels are normally degenerate, so that transitions with $\Delta m = \pm 1$ have no observable consequence. However, in a magnetic induction the sublevels have different energies because of their different magnetic moments $\hbar \gamma m$ or $m \mu_B$ (see eq. (15.62), where for $S = 0$ we have $J = L$ and $g = 1$). Magnetic dipole transitions can therefore occur at an energy $\mu_B B_z$ within these *Zeeman sublevels*. When the potential is not spherically symmetric, the energy eigenstates are no longer simultaneously eigenstates of angular momentum, and allowed magnetic dipole transitions occur even in the absence of a magnetic induction.

A relativistic treatment shows that nuclear and electronic spin can be incorporated in the above arguments by using the complete magnetic dipole moment operator in the matrix element. For atoms the selection rule $\Delta m_J = \pm 1$ is obtained for the z component of the total orbital angular momentum. The magnetic moments of the Zeeman sublevels are then $\hbar\gamma g m_J$ or $g m_J \mu_B$, and allowed magnetic dipole transitions occur at an energy $g\mu_B B_z$ (excluding nuclear spin). In the special case where the orbital angular momentum is zero, we get pure spin states and the usual electron spin resonance transition of energy $2\mu_B B_z$ in the microwave region. Similar considerations apply to nuclear spin states, giving transitions in the radiofrequency region.

The *Zeeman effect* consists of the splitting of spectral lines in a magnetic induction. This is not a magnetic dipole transition within Zeeman sublevels of the same electronic configuration but an electric dipole transition between Zeeman sublevels of different electronic configurations, with $\Delta m = 0$ allowed as well as $\Delta m = \pm 1$. For singlets, with $S = 0$, the *normal* Zeeman effect is observed, with the sublevels split by the same amount in each configuration and hence only three lines in the spectrum. For states with $S \neq 0$, the *anomalous* Zeeman effect is observed: the Landé g-factor and hence the splitting of the sublevels differ in the two configurations, and many more spectral lines result. In either case, interpretation of the spectrum is assisted by the different polarization of the lines according to to the value of Δm.

The *Stark effect* consists of the splitting or shifting of spectral lines in an electric field. The linear Stark effect produces a splitting proportional to the field by lifting a degeneracy, whereas the quadratic Stark effect produces a shift proportional to the square of the field through the usual energy of polarization $-\frac{1}{2}\alpha E^2$. One example of the linear Stark effect occurs in rotational spectroscopy of non-linear molecules, where it allows electric dipole moments to be determined, as described in Section 8.3. A second example occurs in the spectrum of atomic hydrogen, where the degeneracy for a given principal quantum number n is partially lifted. For the case $n = 2$, Example 14.2 shows how the degenerate $2s$ and $2p_z$ states are mixed by a field along the z-axis to yield the states $[|2s\rangle \pm |2p_z\rangle]/2^{1/2}$. These correspond to a movement of electrons with or against the field and hence to a change of energy up or down. A third example occurs in the electronic spectrum of a polar guest molecule which can enter a host molecular crystal lattice in one of two or more equivalent orientations. An electric field lifts the degeneracy of these orientations by interacting with the guest dipole moment in the ground electronic state and in excited electronic states, where the dipole moment is in general different. The guest electronic transitions are therefore split into two or more lines corresponding to sets of molecules differently oriented relative to the field, the splitting depending on the change of dipole moment on excitation. Similarly, the quadratic Stark effect on the electronic spectrum of a guest molecule in a host molecular crystal lattice leads to shifts which depend on the change of polarizability on excitation. In practice these Stark shifts and splittings are small and are measured using a modulated electric field with detection at the appropriate frequency. From such measurements in crystals, information on dipole moment and polarizability changes in excited states can be obtained, subject to the usual need to relate the local field on the molecules to the macroscopic field between the electrodes. A quadratic Stark effect also occurs in the rotational spectroscopy of linear molecules.

16.5 Light scattering

So far we have considered processes in which incident photons stimulate absorption or emission at their own frequency, so that the number of photons changes. It is also found that there are processes in which photons are scattered but their number does not change. The scattering may be *elastic*, with no change of energy, or *inelastic*, with a gain or loss of energy. (Weak higher-order scattering processes with a change of photon number also occur, but these are not considered here.)

Light scattering can be understood as arising from systems in which the electric susceptibility varies in time or space or both. Consider the case of a time-dependent susceptibility

$$\chi(t) = \chi_0 + \chi_t \cos \omega t. \tag{16.32}$$

If the system is subjected to the electric field of a light wave of the form

$$E(t) = E \cos \omega_0 t, \tag{16.33}$$

then the time-dependent polarization at a given point is

$$P(t) = \epsilon_0 \chi(t) E(t) \tag{16.34}$$

$$= \epsilon_0(\chi_0 + \chi_t \cos \omega t) E \cos \omega_0 t. \tag{16.35}$$

Using the trigonometric identity

$$\cos A \cos B = \tfrac{1}{2}[\cos(A+B) + \cos(A-B)], \tag{16.36}$$

we can transform $P(t)$ into

$$P(t) = \epsilon_0 \chi_0 E \cos \omega_0 t + \tfrac{1}{2}\epsilon_0 \chi_t E[\cos(\omega_0+\omega)t + \cos(\omega_0-\omega)t]. \tag{16.37}$$

We therefore have a polarization, and hence a dipole moment, which oscillate not only at the driving frequency of the electric field ω_0, but also at the sum and difference frequencies $\omega_0 \pm \omega$. As we saw in Chapter 6, an oscillating dipole emits radiation at the frequency of oscillation. The incident light wave is thus re-emitted both at its original frequency and at the new frequencies corresponding to the gain or loss of energy $\hbar\omega$ per photon. The elastic scattering is known as *Rayleigh* scattering and the inelastic scattering as *Raman* scattering. The lines in which the incident photon has lost energy are called *Stokes* lines, and those in which the photon has gained energy are called *anti-Stokes* lines.

If the time variation of the susceptibility is not a pure cosine at a single frequency, each Fourier component will give rise to scattering with energy gain or loss at its own frequency. If the susceptibility varies in space instead (or in addition), its spatial variation at a given time combines with that of the electric field of the light wave to give a component of polarization varying with a different wavelength and hence frequency (see Examples, p. 221). This is the mechanism by which dust particles in the air cause scattering to make the sky blue (since high frequencies are scattered most strongly, by eq. (6.60)); by which smoke in fogs and water in mists impair vision; by which nematic liquid crystals, solutions of macromolecules, and materials very close to phase transitions appear turbid or opalescent; and by which density waves in solids and liquids produce *Brillouin* scattering.

In this section we concentrate on Raman scattering caused by a time-dependent susceptibility in molecular systems. The foregoing treatment is purely classical but can be adapted into semi-classical form. The classical scattering intensity expression is used, but the amplitude of the oscillating dipole moment is taken from that of the transition matrix element of the susceptibility change multiplied by the electric vector amplitude E_0. In dilute systems, it suffices to consider the polarizability rather than the susceptibility, the total scattering then being the sum of the molecular contributions. The Raman scattering intensity is governed by matrix elements of the polarizability deviation:

$$\alpha_{\mathrm{lu}} = \langle l \,|\, \alpha(t) - \alpha_0 \,|\, u \rangle, \tag{16.38}$$

where as usual u and l denote the upper and lower molecular states concerned.

Vibrational Raman scattering arises because molecular polarizabilities vary with molecular geometry, which in turn varies during a vibration. An expansion of the polarizability about the equilibrium geometry gives

$$\alpha(x) = \alpha_0 + x(\partial\alpha/\partial x)_0 + \ldots. \tag{16.39}$$

where x is the vibrational displacement, which classically oscillates at the frequency ω of the vibration (assumed harmonic). Terms in x^2 and higher powers of x can normally be neglected. To a good approximation, $(\partial\alpha/\partial x)_0$ depends on the electronic state of the molecule but not on its vibrational state in a given normal mode. The requisite matrix element can thus be written as

$$\alpha_{\mathrm{lu}} = (\partial\alpha/\partial x)_0 \langle l \,|\, \hat{X} \,|\, u \rangle, \tag{16.40}$$

where $\hat{X}$ is the operator corresponding to x.

The selection rules for vibrational Raman scattering follow from the form of α_{lu}. The gross selection rule is that the mode must cause a variation in polarizability, to first order. This is true for symmetric modes, but for antisymmetric modes (in centrosymmetric systems) $(\partial\alpha/\partial x)_0$ is zero, because every change in internuclear separation which tends to increase α is accompanied by an equal and opposite change which tends to decrease it, the net result being zero. So for example in carbon dioxide the symmetric stretch is Raman active but the asymmetric stretch and the bend are not. This is the opposite of the infrared activity, and is an illustration of the *exclusion rule* that in centrosymmetric systems all modes are either infrared active or Raman active but not both or neither. The specific selection rule for vibrational Raman spectra follows from the coordinate matrix element in eq. (16.40), and hence is the same as for infrared spectra governed by electric dipole moment matrix elements, namely $\Delta v = \pm 1$.

Rotational Raman scattering arises because when a molecule rotates its polarizability varies relative to axes fixed in space by the incident light beam. For an axial molecule making an angle θ with the electric vector of the light wave, the polarizability along the electric vector is given by eq. (8.37) as

$$\alpha_\theta = \alpha_{xx} \sin^2\theta + \alpha_{zz} \cos^2\theta, \tag{16.41}$$

where α_{xx} and α_{zz} are the components of polarizability across and along the molecular axis. With the help of trigonometric identities, α_θ can also be written as

$$\alpha_\theta = \tfrac{1}{2}(\alpha_{zz} + \alpha_{xx}) + \tfrac{1}{2}(\alpha_{zz} - \alpha_{xx}) \cos 2\theta. \tag{16.42}$$

where the first term on the right-hand side is the mean polarizability in the molecular plane, corresponding to α_0, and the second term depends on the polarizability anisotropy in the plane. Clearly a molecule with an isotropic polarizability in the plane has the same polarizability in any orientation, by definition, so the variation on rotation must be a function of the anisotropy. Classically, for a free rotor θ is given by ωt, with ω the angular velocity, so that the polarizability deviation varies at *twice* ω. This reflects the fact that as a second-rank tensor the polarizability varies twice as fast as the angle: it regains any value after a rotation of 180° rather than requiring a full rotation of 360°. The matrix element governing scattering in this case is

$$\alpha_{\mathrm{lu}} = \tfrac{1}{2}(\alpha_{zz} - \alpha_{xx})\langle l \,|\cos 2\hat{\Theta}\,|\, u\rangle, \tag{16.43}$$

where $\hat{\Theta}$ is the operator corresponding to θ.

The gross selection rule for rotational Raman scattering is that the polarizability anisotropy in the plane of rotation must be non-zero, as already indicated. The specific selection rules depend on the molecular symmetry rather as for pure rotational absorption spectra. For rigid linear molecules $\Delta J = \pm 2$, while for symmetric tops $\Delta J = \pm 1$, ± 2 and $\Delta K = 0$ (except that for $K = 0$ only $\Delta J = \pm 2$ is allowed); in each case $\Delta J = 0$ is also allowed but corresponds to no change of energy. Rotation about the axis entails no change in polarizability and so is inactive. Asymmetric tops have $\Delta J = 0, \pm 1, \pm 2$, and spherical tops necessarily have no polarizability anisotropy and must be inactive.

An important characteristic of light scattering is that the scattered light from a fluid sample can be polarized even if the incident light is not. The *depolarization ratio* ϱ is defined by

$$\varrho = I_\perp / I_\parallel, \tag{16.44}$$

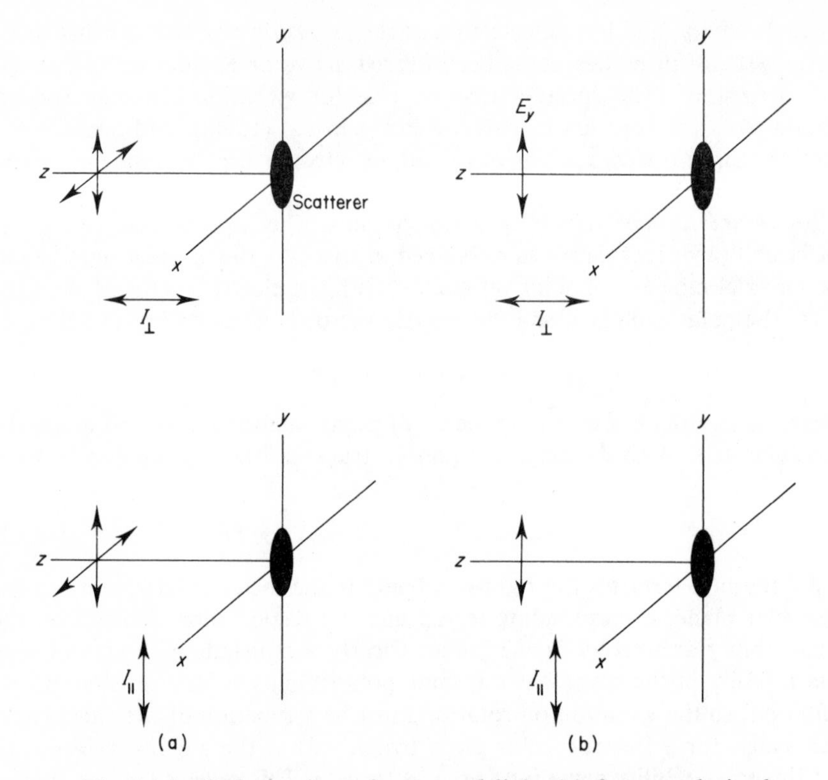

Fig. 16.1 Polarization of scattered light (a) for unpolarized incident light and (b) for polarized incident light, for 90° scattering in each case. The double-headed arrows represent the direction of vibration of the electric vector of the light. The subscripts on the scattered intensity I define the direction of polarization relative to the line or plane of the electric vector

where the scattered intensities $I_\perp$ and $I_\parallel$ are defined by the scattering geometries sketched in Fig. 16.1 for both polarized and unpolarized incident light. To the extent that the light is scattered perpendicular to its original polarization, $I_\perp$ and the depolarization ratio are non-zero.

From eq. (6.60) we know that an oscillating dipole radiates in the plane which contains the dipole and is perpendicular to the direction of oscillation, the light being polarized parallel to the direction of oscillation. Thus in Fig. 16.1, $I_\perp$ arises from an oscillating dipole p_z and $I_\parallel$ from an oscillating dipole p_y, with $I \propto p^2$. These dipoles arise from the oscillating polarizability and incident electric field, and for y-polarized incident light as in Fig. 16.1(b) the dipole amplitudes are

$$P_z = \alpha_{zy} E_y \tag{16.45}$$

$$P_y = \alpha_{yy} E_y, \tag{16.46}$$

where it is to be understood that α_{zy} and α_{yy} here refer to Cartesian components of the transition polarizability tensor matrix element α_{lu}. Hence

$$I_{\perp} \propto \alpha_{zy}{}^2 E_y{}^2 \propto \alpha_{zy}{}^2 I_0 \tag{16.47}$$

$$I_{\parallel} \propto \alpha_{yy}{}^2 E_y{}^2 \propto \alpha_{yy}{}^2 I_0, \tag{16.48}$$

where I_0 is the incident intensity.

The components of polarizability in eqs (16.47) and (16.48) are defined with respect to axes fixed in space by the experimental arrangement in Fig. 16.1. In a fluid sample the molecular axes will be randomly oriented relative to these axes, and the intensities will be determined by averages over all molecular orientations. These averages can be expressed in terms of the principal components α_1, α_2 and α_3 of the transition polarizability and averages of the direction cosines between the space-fixed axes and the molecule-fixed principal axes. The results are found to be

$$\overline{\alpha_{zy}^2} = \gamma^2/15 \tag{16.49}$$

$$\overline{\alpha_{yy}^2} = (45\bar{\alpha}^2 + 4\gamma^2)/45, \tag{16.50}$$

where $\bar{\alpha}$ is the mean or isotropic part of the transition polarizability

$$\bar{\alpha} = \tfrac{1}{3}(\alpha_1 + \alpha_2 + \alpha_3) \tag{16.51}$$

and γ is the anistropy of the transition polarizability, defined by

$$\gamma^2 = \tfrac{1}{2}[(\alpha_1 - \alpha_2)^2 + (\alpha_2 - \alpha_3)^2 + (\alpha_3 - \alpha_1)^2]. \tag{16.52}$$

It follows at once that the depolarization ratio for polarized incident light in the arrangement of Fig. 16.1(b) is

$$\varrho_b = 3\gamma^2/(45\bar{\alpha}^2 + 4\gamma^2). \tag{16.53}$$

For unpolarized incident light in the arrangement of Fig. 16.1(a), a simple extension of the calculation (see Examples, p. 221) gives

$$\varrho_a = 6\gamma^2/(45\bar{\alpha}^2 + 7\gamma^2). \tag{16.54}$$

The depolarization ratios thus lie in the range $0 \leqslant \varrho_b \leqslant 3/4$ and $0 \leqslant \varrho_a \leqslant 6/7$, the lower limit being attained when $\gamma = 0$ and the upper limit when $\bar{\alpha} = 0$. The conditions under which these limits are attained depend on the type of light scattering.

In vibrational Raman scattering, $\bar{\alpha}$ is non-zero only for *totally symmetric* vibrations, i.e. those which preserve the equilibrium molecular geometry (e.g. the symmetric stretch in carbon dioxide). For such modes $\varrho_b < 3/4$ and $\varrho_a < 6/7$, and the line is said to be *polarized*, since it is polarized to more then the minimum extent possble. For spherical top molecules, totally symmetric vibrations give a non-zero $\bar{\alpha}$, but a zero anisotropy γ, so that $\varrho = 0$ and the line is said to be *completely* polarized. Allowed non-totally symmetric vibrations have $\bar{\alpha}$ zero and γ non-zero, so that $\varrho_b = 3/4$ and $\varrho_a = 6/7$, and are said to be *depolarized*, since they have the maximum possible extent of depolarization. In rotational Raman scattering, $\bar{\alpha}$ makes no contribution, as shown by eq. (16.43), and allowed lines are always depolarized. In Rayleigh scattering, however, $\bar{\alpha}$ always contributes and is always non-zero, while γ may be zero (for symmetric tops). The depolarization ratios then lie between zero and 1/3 for ϱ_b or 1/2 for ϱ_a, i.e. less than the usual maximum depolarization ratios.

Measurements of depolarization ratios in rotational Raman and Rayleigh scattering can therefore give the polarizability anisotropy γ if the mean polarizability $\bar{\alpha}$ is known, e.g. from refractive index measurements. Quite subtle information can be obtained this way. For instance, it is found that Rayleigh scattering from the noble gases is not completely polarized, implying that the atoms are not perfectly spherical. This result can be rationalized in terms of atomic collisions which distort the atoms instantaneously from their equilibrium spherical shape.

Light scattering is an important method for studying molecular motions and dimensions via their effects on the coherence of the scattering. This governs the correlation between the electric field amplitude of the scattered radiation $E_s(t)$ at a time t and the amplitude $E_s(0)$ at some arbitrary initial time. The particular information obtained on the correlation depends on the experimental method used, which in turn depends on the time-scale of the process being studied. For processes which occur on a time-scale of μs or faster, filter techniques with diffraction gratings or interferometers are used. These measure the frequency dependence of the field correlations as the Fourier transform of $\langle E_s(0)E_s(t)\rangle$. For slower processes, optical mixing or beating techniques are used: the *homodyne* method studies the scattered light alone and yields information on $\langle |E_s(0)|^2 |E_s(t)|^2\rangle$, while the *heterodyne* method mixes the scattered light with the incident light, yielding information on $\langle E_s(0)E_s(t)\rangle$ instead.

Processing the results of such light-scattering measurements usually depends on solving some model for the fluctuation behaviour of the system. Typically an exponential decay $\langle E_s(t)E_s(0)\rangle \propto e^{-t/\tau}$ is found, where the relaxation time τ depends on the scattering angle and on appropriate properties governing the fluctuations such as the particle diffusion coefficient D. This can yield information on the particle size via the Stokes–Einstein relationship $D = kT/6\pi\eta r$, where η is the viscosity of the solvent and r the radius of the particle, assumed spherical. For non-spherical molecules the motion includes both translational and rotational diffusion, which can be separated by their dependence on scattering angle and analysed to yield the dimensions of the particles, treated as ellipsoids of revolution. For very large molecules, the scattering from different parts of the same molecule need no longer be in phase, so that light scattering on sufficiently dilute solutions of macromolecules can provide information on the relative motion of different segments of the same polymer chain. Very large particles may also give rise to significant phase differences because of local field effects in the interior. Destructive interference of this sort can be minimized by observing the forward scattering, very close to the unscattered beam direction, so that particle sizes can still be determined.

Examples

16.1 Evaluate cos $(\omega r/c)$ for light of wavelength 500 nm at $r = 0$ and $r = 1$ nm.

16.2 Given that $[\hat{A}\hat{B}, \hat{C}] = \hat{A}[\hat{B}, \hat{C}] + [\hat{A}, \hat{C}]\hat{B}$, use eq. (15.31) to show that

$$[\hat{R}_\alpha \hat{R}_\beta, \hat{H}_0] = (i\hbar/m_e)(\hat{R}_\alpha \hat{P}_\beta + \hat{P}_\alpha \hat{R}_\beta),$$

where α and β are Cartesian components. Hence verify eq. (16.19).

16.3 Evaluate the orientationally averaged Einstein coefficient for spontaneous emission for a transition of oscillator strength unity at wavelength of 500 nm. Show that if this were the only decay process the lifetime would be about 11 ns.

16.4 Show that a system with the spatially varying electric susceptibility $\chi(r) = \chi_0 + \chi_r \cos(2\pi r/\lambda)$ gives rise to Stokes and anti-Stokes scattering when subjected to an electric field $E(r) = E \cos(2\pi r/\lambda_0)$.

16.5 Verify eq. (16.54) for the depolarization ratio with unpolarized incident light. Take the components of the electric vector of the light as $E_x = E = E_y$, so that the incident intensity I_0 is proportional to $2E^2$.

Further reading

P. W. Atkins, *Molecular Quantum Mechanics*, 2nd edn, Clarendon Press, Oxford, 1983.

J. L. Martin, *Basic Quantum Mechanics*, Clarendon Press, Oxford, 1981.

P. Landshoff and A. Metherell, *Simple Quantum Physics*, Cambridge University Press, 1979.

J. Avery, *The Quantum Theory of Atoms, Molecules and Photons*, McGraw-Hill, London, 1972.

L. D. Barron, *Molecular Light Scattering and Optical Activity*, Cambridge University Press, 1982.

W. P. Healy, *Non-Relativistic Quantum Electrodynamics*, Academic Press, London, 1982.

D. R. McMillin, Fluctuating electric dipoles and the absorption of light. *J. Chem. Ed.*, **55** (1978) 7.

G. Henderson, How a photon is created or absorbed, *J. Chem Ed.*, **56** (1979) 631.

R. C. Hilborn, Einstein coefficients, cross sections, *f* values, dipole moments, and all that. *Amer. J. Phys.*, **50** (1982) 982.

J. I. Steinfeld, *Molecules and Radiation*, Harper and Row, New York, 1974.

B. J. Berne and R. Pecora, *Dynamic Light Scattering*, Wiley, New York, 1976.

S. Califano, *Vibrational States*, Wiley, London, 1976.

Appendices

Appendix A Vector calculus

A.1 Introduction

We assume that the reader is familiar with the elementary ideas of vector algebra, and simply review salient concepts in this section. We will denote vectors in bold type thus: $\boldsymbol{u}$, $\boldsymbol{v}$, $\boldsymbol{w}$, and unit vectors in the same direction by a 'hat': $\hat{\boldsymbol{u}}$, $\hat{\boldsymbol{v}}$, $\hat{\boldsymbol{w}}$. In the particular case of rectangular Cartesian coordinates we will reserve the symbols $\boldsymbol{i}, \boldsymbol{j}, \boldsymbol{k}$ for unit vectors along the x, y and z axes respectively. Thus, in rectangular Cartesian coordinates a vector is written

$$\boldsymbol{u} = \boldsymbol{i}u_x + \boldsymbol{j}u_y + \boldsymbol{k}u_z$$

where u_x, u_y and u_z are the components of $\boldsymbol{u}$ along the axes (see Fig. A.1).

The *magnitude* of $\boldsymbol{u}$ is

$$u = (u_x^2 + u_y^2 + u_z^2)^{1/2}.$$

The *scalar* (or *dot*) product of the two vectors $\boldsymbol{u}$ and $\boldsymbol{v}$ is defined as

$$\boldsymbol{u} \cdot \boldsymbol{v} = uv \cos\theta = \boldsymbol{v} \cdot \boldsymbol{u}$$

where θ is the angle between $\boldsymbol{u}$ and $\boldsymbol{v}$. In terms of the components of $\boldsymbol{u}$ and $\boldsymbol{v}$,

$$\boldsymbol{u} \cdot \boldsymbol{v} = u_x v_x + u_y v_y + u_z v_z.$$

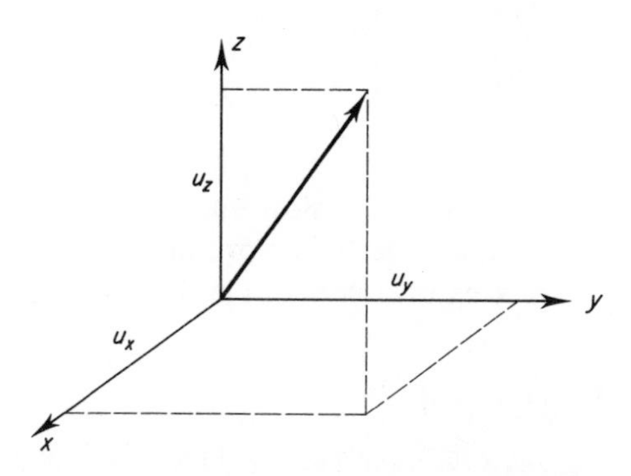

Fig. A.1 Resolution of a vector into three Cartesian rectangular coordinates

The *vector* (or *cross*) product of $\boldsymbol{u}$ and $\boldsymbol{v}$ is written $\boldsymbol{u} \times \boldsymbol{v}$. It is a vector of magnitude $uv \sin \theta$ and direction given by the right-hand screw rule. In terms of components

$$(\boldsymbol{u} \times \boldsymbol{v})_x = u_y v_z - u_z v_y$$

and so on by cyclic permutation of the indices x, y, z. Thus it is seen that $\boldsymbol{u} \times \boldsymbol{v} = -\boldsymbol{v} \times \boldsymbol{u}$. A convenient identity for the evaluation of $\boldsymbol{u} \times \boldsymbol{v}$ is

$$\boldsymbol{u} \times \boldsymbol{v} = \begin{vmatrix} \boldsymbol{i} & \boldsymbol{j} & \boldsymbol{k} \\ u_x & u_y & u_z \\ v_x & v_y & v_z \end{vmatrix}.$$

Having defined the two products $\boldsymbol{u} \cdot \boldsymbol{v}$ and $\boldsymbol{u} \times \boldsymbol{v}$, we can define two different *triple* products. The *scalar triple product* $\boldsymbol{u} \cdot (\boldsymbol{v} \times \boldsymbol{w})$ is formed by first calculating the vector product of $\boldsymbol{v}$ and $\boldsymbol{w}$, and then finding the dot product of this resultant with $\boldsymbol{u}$. In terms of the components of these three vectors it is easily shown that

$$\boldsymbol{u} \cdot (\boldsymbol{v} \times \boldsymbol{w}) = \begin{vmatrix} u_x & u_y & u_z \\ v_x & v_y & v_z \\ w_x & w_y & w_z \end{vmatrix}$$

and it follows from the properties of determinants that

$$\boldsymbol{u} \cdot (\boldsymbol{v} \times \boldsymbol{w}) = \boldsymbol{w} \cdot (\boldsymbol{u} \times \boldsymbol{v}) = \boldsymbol{v} \cdot (\boldsymbol{w} \times \boldsymbol{u}), \text{ etc.}$$

The scalar triple product has the physical significance that it gives the volume of a parallelepiped based on $\boldsymbol{u}$, $\boldsymbol{v}$ and $\boldsymbol{w}$ (see Fig. A.2).

The *vector triple product* $\boldsymbol{u} \times (\boldsymbol{v} \times \boldsymbol{w})$ is formed by first calculating $\boldsymbol{v} \times \boldsymbol{w}$ and then forming the vector product of this resultant with $\boldsymbol{u}$. Unlike the scalar triple product above, the brackets are *essential*, $(\boldsymbol{u} \times \boldsymbol{v}) \times \boldsymbol{w}$ being a quite different vector from $\boldsymbol{u} \times (\boldsymbol{v} \times \boldsymbol{w})$. The vector triple product is found to satisfy

$$\boldsymbol{u} \times (\boldsymbol{v} \times \boldsymbol{w}) = (\boldsymbol{u} \cdot \boldsymbol{w})\boldsymbol{v} - (\boldsymbol{u} \cdot \boldsymbol{v})\boldsymbol{w}$$

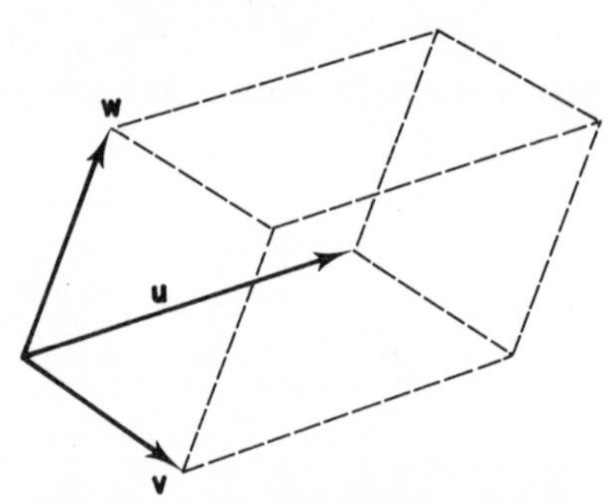

Fig. A.2 Physical interpretation of the scalar triple product as the area of a certain parallepiped

A.2 Scalar and vector fields

Mathematically, a field is a function that describes a physical property at all points in space. In a *scalar field* this physical property is completely described by a single value for each point (e.g. temperature, density, electrostatic potential). For *vector fields* both a value and a direction (or equivalently three components) are required for each point (e.g. gravitational force and electrostatic field intensity).

A.3 Differentiation

Suppose that the vector field $\boldsymbol{u}$ is a continuous function of the scalar variable t. Then as t varies so does $\boldsymbol{u}$ and if $\boldsymbol{u}$ denotes, for example, the position vector of a point P, then P moves along a continuous curve in space as t varies (see Fig. A.3).

By analogy with ordinary calculus, the derivative $\mathrm{d}\boldsymbol{u}/\mathrm{d}t$ is defined as the limit of the ratio $\delta\boldsymbol{u}/\delta t$ as the interval δt becomes progressively smaller. Thus

$$\begin{aligned}\frac{\mathrm{d}\boldsymbol{u}}{\mathrm{d}t} &= \operatorname*{Lim}_{\delta t\to 0} \frac{\delta\boldsymbol{u}}{\delta t} \\ &= \operatorname*{Lim}_{\delta t\to 0}(\boldsymbol{i}\,\delta u_x + \boldsymbol{j}\,\delta u_y + \boldsymbol{k}\,\delta u_z)/\delta t \\ &= \boldsymbol{i}\frac{\mathrm{d}u_x}{\mathrm{d}t} + \boldsymbol{j}\frac{\mathrm{d}u_y}{\mathrm{d}t} + \boldsymbol{k}\frac{\mathrm{d}u_z}{\mathrm{d}t}\end{aligned}$$

and so the derivative of a vector is the vector sum of the derivatives of its components. The usual rules for differentiation are followed:

$$\frac{\mathrm{d}}{\mathrm{d}t}(\boldsymbol{u}+\boldsymbol{v}) = \frac{\mathrm{d}\boldsymbol{u}}{\mathrm{d}t} + \frac{\mathrm{d}\boldsymbol{v}}{\mathrm{d}t}$$

$$\frac{\mathrm{d}}{\mathrm{d}t}(a\boldsymbol{u}) = a\frac{\mathrm{d}\boldsymbol{u}}{\mathrm{d}t} \qquad (a = \text{scalar constant})$$

$$\frac{\mathrm{d}}{\mathrm{d}t}(f\boldsymbol{u}) = \frac{\mathrm{d}f}{\mathrm{d}t}\boldsymbol{u} + f\frac{\mathrm{d}\boldsymbol{u}}{\mathrm{d}t} \qquad (f = \text{scalar field})$$

$$\frac{\mathrm{d}}{\mathrm{d}t}(\boldsymbol{u}\cdot\boldsymbol{v}) = \boldsymbol{u}\cdot\frac{\mathrm{d}\boldsymbol{v}}{\mathrm{d}t} + \boldsymbol{v}\cdot\frac{\mathrm{d}\boldsymbol{u}}{\mathrm{d}t}$$

$$\frac{\mathrm{d}}{\mathrm{d}t}(\boldsymbol{u}\times\boldsymbol{v}) = \boldsymbol{u}\times\frac{\mathrm{d}\boldsymbol{v}}{\mathrm{d}t} + \frac{\mathrm{d}\boldsymbol{u}}{\mathrm{d}t}\times\boldsymbol{v}.$$

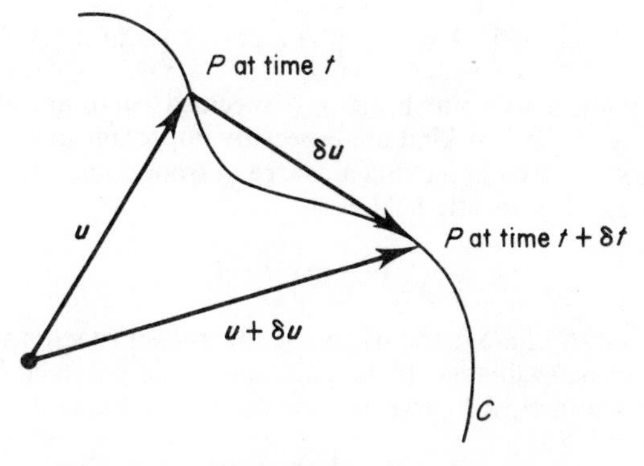

Fig. A.3 Illustrating the idea of the differential with respect to the scalar variable t of the vector field $\boldsymbol{u}$

A.4 The gradient

We shall be interested in one particular function of the spatial derivatives of a scalar field and in two particular spatial derivatives of a vector field. All these derivatives turn out to be useful in electromagnetism.

Suppose that $f(x, y, z)$ is a scalar field and we wish to know how f changes between the point (x, y, z) and a point displaced by the infinitesimal vector

$$\mathrm{d}\boldsymbol{l} = \boldsymbol{i}\,\mathrm{d}x + \boldsymbol{j}\,\mathrm{d}y + \boldsymbol{k}\,\mathrm{d}z.$$

We know from elementary calculus that

$$\mathrm{d}f = \left(\frac{\partial f}{\partial x}\right)\mathrm{d}x + \left(\frac{\partial f}{\partial y}\right)\mathrm{d}y + \left(\frac{\partial f}{\partial z}\right)\mathrm{d}z$$

where, for example, $(\partial f/\partial x)$ implies that y and z are held constant during the differentiation, and we can write $\mathrm{d}f$ as the scalar product of a certain vector $\boldsymbol{A}$ with $\mathrm{d}\boldsymbol{l}$:

$$\mathrm{d}f = \boldsymbol{A}\cdot\mathrm{d}\boldsymbol{l},$$

where by inspection

$$\boldsymbol{A} = \frac{\partial f}{\partial x}\boldsymbol{i} + \frac{\partial f}{\partial y}\boldsymbol{j} + \frac{\partial f}{\partial z}\boldsymbol{k}.$$

This vector $\boldsymbol{A}$ is called the *gradient of f* and is written ∇f where the gradient operator ∇ (or 'del') is given by

$$\nabla \equiv \boldsymbol{i}\frac{\partial}{\partial x} + \boldsymbol{j}\frac{\partial}{\partial y} + \boldsymbol{k}\frac{\partial}{\partial z}.$$

The magnitude of $\mathrm{d}f$ can also be written

$$\mathrm{d}f = \nabla f\cdot\mathrm{d}\boldsymbol{l}$$

from which we deduce that ∇f is a vector whose magnitude and direction are those of the maximum spatial rate of change of f. Thus ∇f points towards *larger* values of f.

A.5 Line integrals

The integrals

$$\int_a^b \boldsymbol{A}\cdot\mathrm{d}\boldsymbol{l} \qquad \int_a^b \boldsymbol{A}\times\mathrm{d}\boldsymbol{l} \qquad \int_a^b f\,\mathrm{d}\boldsymbol{l}$$

evaluated from point a to point b along a specified curve are all examples of line integrals. Integrals of the first kind are especially important in electromagnetism; for example, the work involved in moving a charge Q from point a to point b along some specified path in an electrostatic field $\boldsymbol{E}$ is

$$W = Q\int_a^b \boldsymbol{E}\cdot\mathrm{d}\boldsymbol{l}$$

Both $\boldsymbol{E}$ and the path (i.e. $\mathrm{d}\boldsymbol{l}$) must, of course, be known functions of the coordinates if the integral is to be evaluated. If the final and initial points a, b coincide then the line integral refers to a closed curve and we denote the integral

$$\oint \boldsymbol{A}\cdot\mathrm{d}\boldsymbol{l}$$

If the line integral of a certain vector field $\boldsymbol{A}$ is zero around *any* (i.e. arbitrary) closed curve, then the vector field is called a *conservative* field.

A.6 Surface and volume integrals

Consider a vector field A and a surface S (Fig. A.4).

We divide the surface into small elements $\delta S_1, \ldots, \delta S_n$ as shown such that δS_k is the area of each subdivision. Suppose that the (average) values of A calculated within each δS are $A_1, \ldots, A_k$ respectively and let the vectors $n_1, \ldots, n_k$ represent unit normals to each of $\delta S_1, \ldots, \delta S_k$. The direction of each n_k is conventionally taken to be the *outward* normal. If we calculate

$$\sum A_i \cdot n_i \, dS_i$$

and let the δS_k become infinitesimally small $\delta S \rightarrow dS$, then this expression defines a *surface integral*

$$\lim_{\delta S_i \rightarrow 0} \sum A_i \cdot n_i \, \delta S_i = \int A \cdot n \, dS.$$

It is usual to define a *surface element*

$$d\mathbf{S} = n \, dS$$

as a vector whose magnitude is equal to the area dS and whose direction is that of the outward normal. Thus the surface integral is written

$$\int A \cdot d\mathbf{S}$$

and if the surface S is a closed surface

$$\oint A \cdot d\mathbf{S},$$

where $d\mathbf{S}$ is everywhere directed outwards. Clearly, in order to *evaluate* such an integral we need to be able to express A and $d\mathbf{S}$ in some functional form. An integral evaluated over two variables (that are not necessarily spatial variables) is called a double integral and methods for evaluating double integrals are discussed in standard calculus texts. The integral $\int A \cdot d\mathbf{S}$ is clearly such a double integral.

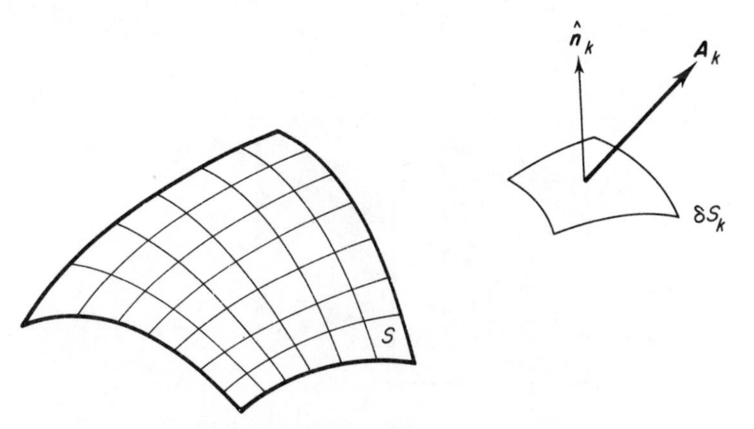

Fig. A.4 Division of an arbitrary surface S into infinitesimal surface elements δS_k. The flux of a vector field A through δS_k may not be parallel to δS_k; n_k is a unit normal

In a *volume integral*

$$\int_V A \, d\tau$$

$d\tau$ is a volume element of some region of space V. This is a triple integral (e.g. with $d\tau = dx\,dy\,dz$) and again methods of evaluating triple integrals are discussed in standard texts.

A.7 Flux

In electromagnetism it is often necessary to calculate the *flux* of a vector field through a surface. The flux of A through the surface element dS is $A \cdot dS$, and for finite surfaces we integrate:

$$\Phi = \int_s A \cdot dS$$

with the familiar sign convention for the direction of dS.

A.8 The divergence theorem (Gauss' theorem)

We saw from Section A.7 above that the flux of a vector field A through the surface S can be calculated as

$$\int_s A \cdot dS.$$

It turns out, however, that there is an alternative method, as follows. Consider the infinitesimal volume element $dx\,dy\,dz$ centred on the point (x, y, z) as shown in Fig. A.5.

Let the vector field A have components which are functions of the coordinates x, y, z, and consider the flux of A through the two shaded faces. The value of A_x at the right-hand face is the average value over that face, so the *outgoing* flux is

$$d\Phi_R = \left(A_x + \frac{\partial A_x}{\partial x}\frac{dx}{2}\right) dy\,dz.$$

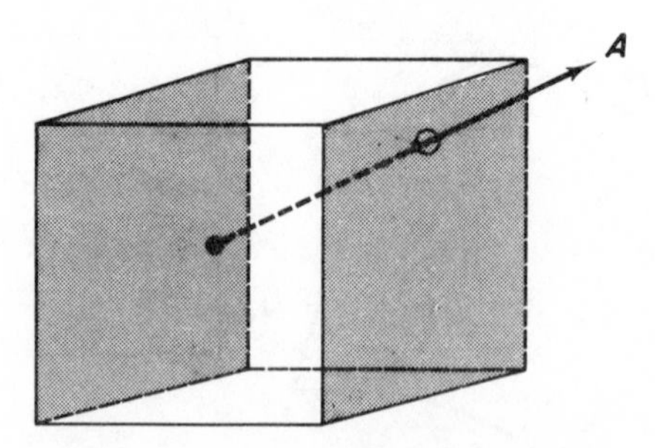

Fig. A.5 Calculation of the flux of **A** through a differential volume element, from which the divergence theorem follows

At the left-hand face the flux is *inward*

$$d\Phi_L = -\left(A_x - \frac{\partial A_x}{\partial x}\frac{dx}{2}\right) dy\, dz$$

so that the total outward flux through these two faces is $(\partial A_x/\partial x)\, dx\, dy\, dz$. If we calculate the total flux through all faces of the cube and add, we find

$$\Phi = \oint_S \left(\frac{\partial A_x}{\partial x} + \frac{\partial A_y}{\partial y} + \frac{\partial A_z}{\partial z}\right) dx\, dy\, dz$$

At any given point in the volume, the quantity

$$\frac{\partial A_x}{\partial x} + \frac{\partial A_y}{\partial y} + \frac{\partial A_z}{\partial z}$$

gives the total outgoing flux per unit volume and we call this the *divergence* of the vector field $\boldsymbol{A}$ at the point. This may be written as div $\boldsymbol{A}$ but is more informatively written in terms of the *gradient operator* as

$$\nabla \cdot \boldsymbol{A} = \frac{\partial A_x}{\partial x} + \frac{\partial A_y}{\partial y} + \frac{\partial A_z}{\partial z}$$

which shows clearly that the divergence is a scalar.

To extend this calculation to a finite volume instead of the infinitesimal element of Fig. A.5 it is necessary to sum the individual fluxes of a large number of infinitesimal elements and when this is done we find

$$\Phi_{tot} = \int \nabla \cdot \boldsymbol{A}\, d\tau$$

where we have written $d\tau$ for the volume element $dx\, dy\, dz$. But from Section A.7 this flux is $\oint \boldsymbol{A} \cdot d\boldsymbol{S}$ and hence

$$\int_V \nabla \cdot \boldsymbol{A}\, d\tau = \oint_S \boldsymbol{A} \cdot d\boldsymbol{S}$$

which is the *divergence theorem*. Note that the integral on the right-hand side involves values of $\boldsymbol{A}$ on the surface *only* whereas the left-hand integral involves values of the divergence of $\boldsymbol{A}$ throughout the volume. (The reason for this is that when individual infinitesimal fluxes are added, there is cancellation between adjacent infinitesimal volume elements.)

A.9 The curl

We discussed earlier the concept of a line integral $\int \boldsymbol{A} \cdot d\boldsymbol{l}$ and introduced the idea of a conservative vector field, one for which $\oint \boldsymbol{A} \cdot d\boldsymbol{l} = 0$ around any closed path. Let us calculate the value of the integral in a more general case. For simplicity we will treat an infinitesimal closed path dC in the xy plane as in Fig. A.6.

$$\oint_{dC} \boldsymbol{A} \cdot d\boldsymbol{l} = \oint A_x\, d_x + \oint A_y\, dy$$

and following a similar reasoning to Section A.8, we can evaluate both integrals on the right-hand side;

$$\oint_{dC} A_x\, dx = \left(A_x - \frac{\partial A_x}{\partial x} \cdot \frac{dy}{2}\right) dx - \left(A_x + \frac{\partial A_x}{\partial y} \cdot \frac{dy}{2}\right)$$

with the negative sign between the two brackets arising because the path element at

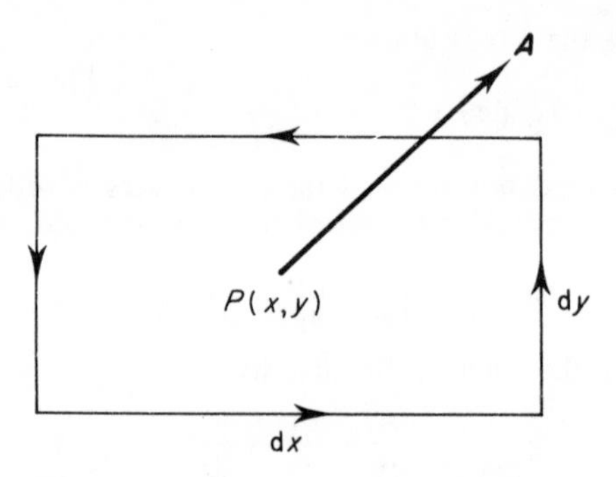

Fig. A.6 Figure for the proof of Stokes' theorem

$y+\mathrm{d}y/2$ is in the negative x direction. Thus

$$\oint_{\mathrm{d}C} A_x\,\mathrm{d}x = -\frac{\partial A_x}{\partial y}\cdot\mathrm{d}y\,\mathrm{d}x$$

and similarly

$$\oint_{\mathrm{d}C} A_y\,\mathrm{d}y = \frac{\partial A_y}{\partial x}\cdot\mathrm{d}x\,\mathrm{d}y$$

Thus, adding,

$$\oint_{\mathrm{d}C} \boldsymbol{A}\cdot\mathrm{d}\boldsymbol{l} = \left(\frac{\partial A_y}{\partial x} - \frac{\partial A_x}{\partial y}\right)\mathrm{d}x\,\mathrm{d}y.$$

(Note that this is only correct if the line integral is calculated along the path according to the right-hand screw rule, i.e. the direction one would have to turn a right-hand screw in order to make it advance in a positive direction along the z-axis. This is the standard convention for line integrals.)

The quantity $\partial A_y/\partial x - \partial A_x/\partial y$ is the z component of the vector $\nabla\times\boldsymbol{A}$, called the *curl* of $\boldsymbol{A}$.

For a *general* infinitesimal line integral where the rectangle of Fig. A.6 does not lie in the xy plane, it turns out that the result above becomes

$$\oint_{\mathrm{d}C} \boldsymbol{A}\cdot\mathrm{d}\boldsymbol{l} = (\nabla\times\boldsymbol{A})\cdot\mathrm{d}\boldsymbol{S}$$

where the surface element $\mathrm{d}\boldsymbol{S}$ is infinitesimal. If this result is added up for all the elements $\mathrm{d}\boldsymbol{S}$ making up a finite surface S, the infinitesimal paths $\mathrm{d}C$ give contributions which are cancelled by equal and opposite contributions from adjacent paths, except along the path C bounding the surface S. We then obtain a relationship between a finite line integral and a finite surface integral:

$$\oint_C \boldsymbol{A}\cdot\mathrm{d}\boldsymbol{l} = \int_S (\nabla\times\boldsymbol{A})\cdot\mathrm{d}\boldsymbol{S}$$

and this result is known as *Stokes' theorem*.

A.10 The Laplacian

It turns out that the divergence of the gradient of certain fields are important in electromagnetism (as in science and engineering generally). Since

$$\nabla\phi = \frac{\partial\phi}{\partial x}\boldsymbol{i} + \frac{\partial\phi}{\partial y}\boldsymbol{j} + \frac{\partial\phi}{\partial z}\boldsymbol{k}$$

then

$$\nabla\cdot(\nabla\phi) \equiv \nabla^2\phi = \frac{\partial^2\phi}{\partial x^2} + \frac{\partial^2\phi}{\partial y^2} + \frac{\partial^2\phi}{\partial z^2}$$

in rectangular cartesian coordinates; ∇^2 is called the *laplacian*. It is often necessary to work in coordinate systems other than rectangular Cartesian ones. For convenience various differential expressions are summarized in Section A.12 in both rectangular Cartesian and spherical polar coordinates. Transformation to other coordinate systems is straightforward but tedious.

A.11 Conservative fields

We have already encountered the idea of a conservative field, one for which

$$\oint_C \boldsymbol{A}\cdot d\boldsymbol{l} = 0$$

for every closed path C. We can immediately invoke Stokes' theorem

$$\oint_C \boldsymbol{A}\cdot d\boldsymbol{l} = \int_S (\nabla\times\boldsymbol{A})\cdot d\boldsymbol{S} = 0$$

and because we have made no use of any special path C (it is arbitrary) we deduce that the right-hand integrand is zero: $\nabla\times\boldsymbol{A} = \boldsymbol{0}$. Thus for a conservative field $\boldsymbol{A}$, $\nabla\times\boldsymbol{A} = \boldsymbol{0}$.

Now, $\nabla\times(\nabla\phi) = \boldsymbol{0}$ for any differentiable scalar field: this is a vector identity easily proved by expanding the left-hand side into components. Hence, for a conservative field $\boldsymbol{A}$, we can represent $\boldsymbol{A}$ as the gradient of a scalar field $\boldsymbol{A} = \nabla\phi$ and thereby automatically ensure that $\nabla\times\boldsymbol{A} = \boldsymbol{0}$. Note that ϕ is indeterminate to within a constant. Thus, for a conservative field A all of the following are valid:

(1) $\oint_C \boldsymbol{A}\cdot d\boldsymbol{l} = 0$ for any closed path C.

(2) $\nabla\times\boldsymbol{A} = \boldsymbol{0}$ everywhere.

(3) We can find a scalar field ϕ such that $\boldsymbol{A} = \nabla\phi$.

A.12 Vector identities

The following vector identities are used in the text. All may be proved by expanding right- and left-hand sides and comparing components; $\boldsymbol{A}$ and $\boldsymbol{B}$ are vector fields, V and W are scalar fields.

(1) $\nabla\cdot(\boldsymbol{A}+\boldsymbol{B}) \equiv \nabla\cdot\boldsymbol{A} + \nabla\cdot\boldsymbol{B}$

(2) $\nabla(V+W) \equiv \nabla V + \nabla W$

(3) $\nabla\times(\boldsymbol{A}+\boldsymbol{B}) \equiv \nabla\times\boldsymbol{A} + \nabla\times\boldsymbol{B}$

(4) $\nabla\cdot(V\boldsymbol{A}) \equiv \boldsymbol{A}\cdot\nabla V + V\nabla\cdot\boldsymbol{A}$

(5) $\nabla(VW) \equiv V\nabla W + W\nabla V$

(6) $\nabla \times (V\boldsymbol{A}) \equiv \nabla V \times \boldsymbol{A} + V\nabla \times \boldsymbol{A}$

(7) $\nabla \cdot (\boldsymbol{A} \times \boldsymbol{B}) \equiv \boldsymbol{B} \cdot \nabla \times \boldsymbol{A} - \boldsymbol{A} \cdot \nabla \times \boldsymbol{B}$

(8) $\nabla (\boldsymbol{A} \cdot \boldsymbol{B}) \equiv (\boldsymbol{A} \cdot \nabla)\boldsymbol{B} + (\boldsymbol{B} \cdot \nabla)\boldsymbol{A} + \boldsymbol{A} \times (\nabla \times \boldsymbol{B}) + \boldsymbol{B} \times (\nabla \times \boldsymbol{A})$

(9) $\nabla \times (\boldsymbol{A} \times \boldsymbol{B}) \equiv \boldsymbol{A}\nabla \cdot \boldsymbol{B} - \boldsymbol{B}\nabla \cdot \boldsymbol{A} + (\boldsymbol{B} \cdot \nabla)\boldsymbol{A} - (\boldsymbol{A} \cdot \nabla)\boldsymbol{B}$

(10) $\nabla \cdot \nabla \times \boldsymbol{A} \equiv 0$

(11) $\nabla \times \nabla V \equiv \boldsymbol{0}$

(12) $\nabla \times (\nabla \times \boldsymbol{A}) \equiv \nabla (\nabla \cdot \boldsymbol{A}) - \nabla^2 \boldsymbol{A}$

A.13 Differential expressions

(a) Rectangular Cartesian coordinates

$$\nabla V = \boldsymbol{i}\frac{\partial V}{\partial x} + \boldsymbol{j}\frac{\partial V}{\partial y} + \boldsymbol{k}\frac{\partial V}{\partial z}$$

$$\nabla \cdot \boldsymbol{A} = \frac{\partial A_x}{\partial x} + \frac{\partial A_y}{\partial y} + \frac{\partial A_z}{\partial z}$$

$$\nabla \times \boldsymbol{A} = \boldsymbol{i}\left(\frac{\partial A_y}{\partial z} - \frac{\partial A_z}{\partial y}\right) + \boldsymbol{j}\left(\frac{\partial A_z}{\partial x} - \frac{\partial A_x}{\partial z}\right) + \boldsymbol{k}\left(\frac{\partial A_x}{\partial y} - \frac{\partial A_y}{\partial x}\right)$$

$$\nabla^2 V = \frac{\partial^2 V}{\partial x^2} + \frac{\partial^2 V}{\partial y^2} + \frac{\partial^2 V}{\partial z^2}$$

Volume element $d\tau = dx \; dy \; dz$

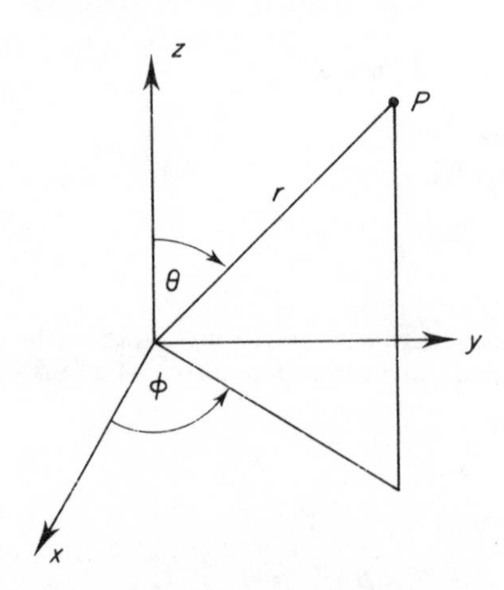

Fig. A.7 Connection between rectangular Cartesian and spherical polar coordinates

(b) Spherical polar coordinates

$$\nabla V = \hat{r}\frac{\partial V}{\partial r} + \hat{\theta}\frac{1}{r}\frac{\partial V}{\partial \theta} + \hat{\phi}\frac{1}{r\sin\theta}\frac{\partial V}{\partial \phi}$$

$$\nabla \cdot A = \frac{1}{r^2}\frac{\partial}{\partial r}(r^2 A_r) + \frac{1}{r\sin\theta}\frac{\partial}{\partial \theta}(A_\theta \sin\theta) + \frac{1}{r\sin\theta}\frac{\partial A_\phi}{\partial \phi}$$

$$\nabla \times A = \hat{r}\frac{1}{r\sin\theta}\left[\frac{\partial}{\partial \theta}(\sin\theta\, A_\theta) - \frac{\partial A_\theta}{\partial \phi}\right] + \hat{\theta}\frac{1}{r}\left[\frac{1}{\sin\theta}\frac{\partial A_r}{\partial \phi} - \frac{\partial}{\partial r}(rA_\phi)\right] + \hat{\phi}\frac{1}{r}\left[\frac{\partial}{\partial r}(rA_\theta) - \frac{\partial A_r}{\partial \theta}\right]$$

$$\nabla^2 V = \frac{1}{r^2}\frac{\partial}{\partial r}\left(r^2\frac{\partial V}{\partial r}\right) + \frac{1}{r^2\sin\theta}\frac{\partial}{\partial \theta}\left(\sin\theta\frac{\partial V}{\partial \theta}\right) + \frac{1}{r^2\sin^2\theta}\frac{\partial^2 V}{\partial \phi^2}$$

Volume element $d\tau = r^2 \sin\theta \, dr \, d\theta \, d\phi$

A.14 Tensors

In a number of places, quantities are referred to as *tensors*. We make no use of formal tensor analysis in this book, but this section gives some background and suggestions for further reading.

The quantities which we know as vectors have both magnitude and direction. They can therefore be conveniently specified by three components relative to some orthogonal set of Cartesian axes, as we have seen. The choice of axes is not fixed, and so the same vector can be specified by different sets of components relative to different sets of axes. The transformation between the components in different sets of axes then depends linearly on the direction cosines between the different sets of axes; it has the same algebraic form for all vector quantities, and so can be taken as defining a vector.

Tensors describe other quantities which have related but more complicated properties. For example, the nine various products of the components of a pair of vectors will transform between different sets of axes with coefficients which depend on products of direction cosines. The quadrupole moment is a quantity of this type. Alternatively, the components of some vector response of a system and those of the vector stimulus producing the response are related by a set of nine coefficients which transform between axes in the same way depending on products of direction cosines. An example here is the polarizability. All quantities with the transformation properties of a product of two vectors are known as (second-rank) tensors. There are also tensors of higher ranks r such that the components transform like those of a product of r vectors.

The advantage of introducing tensors is that they simplify the algebraic forms of expressions much as vectors do. Thus the relation between the induced dipole moment $\boldsymbol{p}$ and the electric field $\boldsymbol{E}$ via the polarizability tensor $\boldsymbol{\alpha}$ can be written as

$$\boldsymbol{p} = \boldsymbol{\alpha} \cdot \boldsymbol{E}$$

meaning that

$$p_i = \sum_{j=1}^{3} \alpha_{ij} E_j,$$

where $i = 1, 2,$ or 3 labels the axis along which the component is taken. Similarly the energy required to induce the dipole moment can be written as

$$W = -\tfrac{1}{2}\boldsymbol{E} \cdot \boldsymbol{\alpha} \cdot \boldsymbol{E}.$$

meaning

$$W = -\frac{1}{2} \sum_{ij} E_i \alpha_{ij} E_j.$$

Tensors can also be combined together, as in the energy of a quadrupole moment $\mathbf{Q}$ in an electric field gradient $\mathbf{E}'$, which is written as

$$W = -\frac{1}{2} \mathbf{Q} : \mathbf{E}'$$

meaning

$$W = -\frac{1}{2} \sum_{ij} Q_{ij} E'_{ij}.$$

Detailed accounts of tensors may be found in *Cartesian Tensors* by H. Jeffreys (Cambridge University Press, 1969) and *Cartesian Tensors: An Introduction* by G. Temple (Wiley, New York, 1960).

Appendix B Useful quantities

Physical constants

Constant	Symbol	Value	
Gravitational constant	G	6.672×10^{-11}	$\mathrm{N\,m^2\,kg^{-2}}$
Avogadro constant	L	6.022×10^{23}	$\mathrm{mol^{-1}}$
Proton rest mass	m_p	1.673×10^{-27}	kg
Electron rest mass	m_e	9.109×10^{-31}	kg
Elementary charge	e	1.602×10^{-19}	C
Permittivity of free space	ϵ_0	8.854×10^{-12}	$\mathrm{F\,m^{-1}}$
Permeability of free space	μ_0	$4\pi \times 10^{-7}$	$\mathrm{H\,m^{-1}}$ (exactly)
Speed of light in free space	c_0	2.998×10^{8}	$\mathrm{m\,s^{-1}}$
Planck constant	h	6.626×10^{-34}	J s
Boltzmann constant	k	1.381×10^{-23}	$\mathrm{J\,K^{-1}}$
Bohr magneton	μ_B	9.274×10^{-24}	$\mathrm{A\,m^2}$ (or $\mathrm{J\,T^{-1}}$)

Electromagnetic quantities

Name	Symbol	Unit
Force	F	N
Charge	Q	C
Electric field intensity	E	$V\,m^{-1}$
Volume charge density	ϱ	$C\,m^{-3}$
Surface charge density	σ	$C\,m^{-2}$
Line charge density	μ	$C\,m^{-1}$
Work, energy	W	J
Scalar potential	V	V
Electric dipole moment	p_e	C m
Electric quadrupole moment	Q_e	$C\,m^2$
Current	I	A
Current density	J	$A\,m^{-2}$
Conductivity	$\varkappa$	$S\,m^{-1}$
Polarization	P	$C\,m^{-2}$
Electric susceptibility	χ_e	(number)
Capacitance	C	F
Magnetic field intensity	H	$A\,m^{-1}$
Magnetic flux density	B	$Wb\,m^{-2}$ (or T)
Magnetic flux	Φ	Wb
Vector potential	A	$Wb\,m^{-1}$
Magnetic moment	p_m or m	$A\,m^2$ (or $J\,T^{-1}$)
Magnetization	M	$A\,m^{-1}$
Magnetic susceptibility	χ_m	(number)
Inductance	L	H
Poynting vector	S	$W\,m^{-2}$

Solutions and hints to examples

1.4 If the point charge is at the centre of the ring, the resultant force is zero. For axial points, resolve the force due to a differential element of the ring into a parallel and a perpendicular component. The parallel component will be cancelled by a differential element 180° further round the ring. 5.13×10^6 N.

1.5 $Q_3 = -4Q_1$

1.7 This example is somewhat artificial in treating an *infinite* sheet of charge, and the potential at infinity cannot therefore be taken to be zero. In any case $\boldsymbol{E} = -\nabla V$ so write V_∞ for the arbitrary value of V at infinity, whence $V = V_\infty - a\sigma/2\epsilon_0$ where a is the scalar distance from the field point to the sheet.

1.9 Change the coordinate origin and examine the relationship between the definition of $\boldsymbol{p}$ in both coordinate systems.

1.10 The charge enclosed by a spherical gaussian surface centred on the nucleus is

$$e\left(1 - \int_0^R \Psi^2 \, \mathrm{d}\tau\right)$$

where Ψ is the electronic wavefunction. The volume element $\mathrm{d}\tau = 4\pi r^2 dr$ and the integral

$$\int_0^a x^2 \exp(-bx) \, \mathrm{d}x$$

is available in any standard compilation of integrals.

1.11 $A = 3Q/\pi R^4$. At distances where $r > R$

$$E_n = Q/4\pi\epsilon_0 r^2.$$

For $r \leqslant R$ the enclosed charge is

$$\frac{Qr^3}{R^4}\left(4R - 3r\right)$$

and the electrostatic field is

$$E_n = \frac{Q}{4\pi\epsilon_0} \frac{r(4R - 3r)}{R^4}$$

1.12 $\varrho_0 = 15Q/8\pi a^2$. For points $r \geqslant a$

$$E_n = 2\varrho_0 a^3/15\epsilon_0 r^2$$

and for points $r \leqslant a$ the enclosed charge is

$$4\pi\varrho_0\left[\frac{r^3}{3} - \frac{r^5}{5a^2}\right]$$

and hence

$$E_n = \frac{\varrho_0 r}{\epsilon_0}\left[\frac{1}{3} - \frac{r^2}{5a^2}\right]$$

1.13 A conservative field A is one for which $\nabla \times A = 0$ and $\oint_C A \cdot dl = 0$ for any closed path C. All central fields (those which only depend on the scalar distance from the origin) are conservative fields. This can be shown either by calculating their curl or by evaluation of a closed line integral. The flux of A through a sphere of raduis r centred at the origin would be $\int E \cdot dS \propto r^{k+1} 4\pi r^2$ which is independent of r only if $k = -3$

1.14

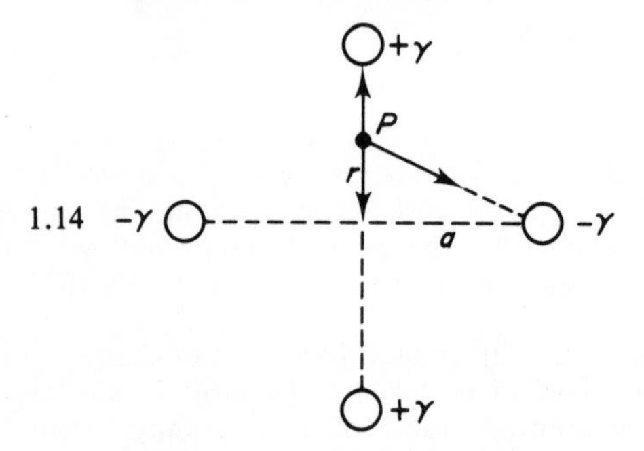

Consider the point P distance r from the centre and add together the forces shown. $E_{tot} = 2\gamma r/\pi\epsilon_0 a^2$ and hence

$$\frac{\partial E}{\partial r} = 2\gamma/\pi\epsilon_0 a^2$$

2.1 The corresponding result is the negative of the one quoted.

2.4 $p = p' = 5.97 \times 10^{-30}$ Cm and so $F = 2.37 \times 10^{-10}$ N.

2.5 From the text $\varrho_b = -\nabla \cdot P$ and $\sigma_b = P \cdot n$. Substitute and use the divergence theorem.

2.6 Total energy $U = Q^2/8\pi\epsilon_0 r + 4\pi r^2 \gamma$ and as r^2 increases so $1/r$ decreases. There will be a balance when $dU/dr = 0$ and this establishes the result.

2.7 It is probably easier to work with D rather than E, and we have to assume that the media are linear and isotropic.

$$\nabla \cdot D = \varrho_{free} \qquad \epsilon E = D$$

Calculate the potential

$$-V = \left\{\int_{\infty}^{r_3} + \int_{r_3}^{r_2} + \int_{r_2}^{r_1}\right\} D \cdot dl$$

3.1 The direction of B is that of $dl \times R$ and is therefore perpendicular to the loop.

3.2

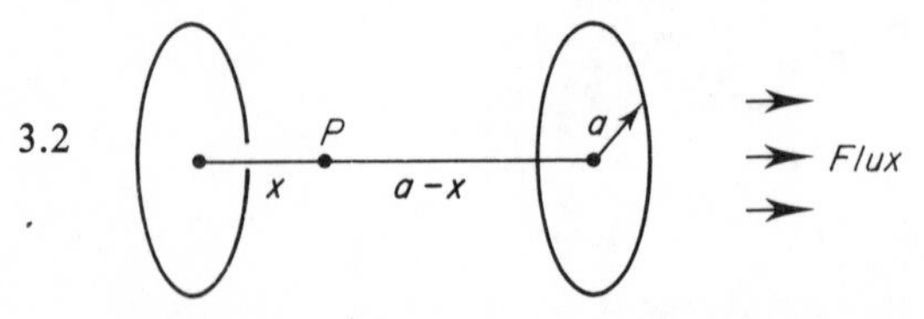

Establish that the magnetic induction along the principal axis of a current loop is

$$\hat{n}\mu_0 I a^2/2(a^2 + x^2)^{3/2}$$

and add together the values for both loops at point P. Find the first and second differential with respect to x and evaluate these quantities when $x = a/2$.

3.3 Split the solenoid into loops of radius a and width dx. Thus integrate with respect to x.

3.4 As Example 3.3, but think about the limits of integration.

3.5 Use Ampère's circuital law with a path shown. The enclosed current is $I \sin(\pi r/a)$ and hence

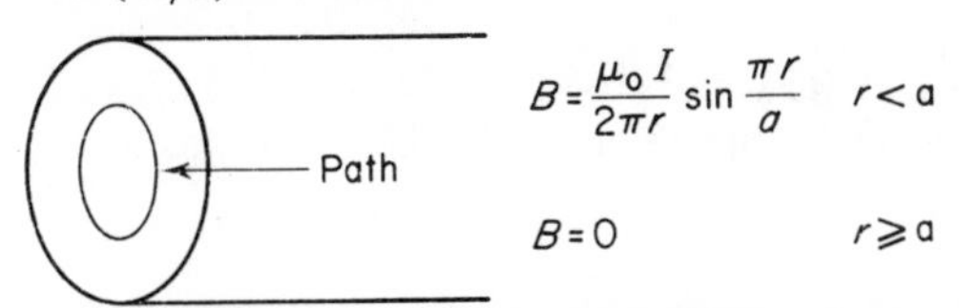

$$B = \frac{\mu_0 I}{2\pi r} \sin \frac{\pi r}{a} \qquad r < a$$

$$B = 0 \qquad r \geqslant a$$

3.6 $v \times B = B(iv_y - jv_x)$. Write down the Lorentz force and separate for x and y. To solve, differentiate again and substitute for v_x, v_y.

3.7 $v = 4.2 \times 10^7 \text{ m s}^{-1}$, $B = 4.75 \times 10^{-3} \text{ Wb m}^{-2}$.

3.9 $E = (E, 0, 0)$ and $B = (0, 0, B)$ Write down the Lorentz force and separate into three components. Integrate the y equation and substitute into the x equation. Integrate, giving

$$x = \frac{Em}{eB^2}\left(1 - \cos\frac{eB}{m}t\right)$$

Hence the result

3.11 This is straightforward but because the problem is artificial in having an 'infinite' wire there is a difficulty in the constant of integration.

5.1 The force is everywhere tangential to the ring and

$$F_t = \frac{e}{2\pi r}\frac{d\Phi}{dt} = \text{mass} \times \text{acceleration} = mr\frac{d\omega}{dt}$$

The flux through the benzene 'ring' (!) of radius a is $B_0 \pi a^2$. Integrate and substitute.

$$\text{Field at centre} = -\frac{6e^2 B\mu_0}{m}$$

6.2 Take the divergence of the curl E equation to derive the first result. Start from $\nabla \times E = -J^*$, whence $\int(\nabla \times E)\cdot dS = -\int J^* \cdot dS$ for any surface S, and use Stokes' theorem to obtain the second result.

6.3 Substitute E' and B' into Maxwell's equations and verify that they do hold. Maxwell's equations are linear in the fields E and B, so in general linear combinations of the fields will satisfy the equations, provided that the fields are themselves solutions.

6.4 We have $\nabla \times E = -\mu_0 \partial H/\partial t$ with $c^2 = (\mu_0\epsilon_0)^{-1}$ Substitute, calculate $\nabla \times E$ and integrate.

$$H = -j\left(\frac{\epsilon_0}{\mu_0}\right)^{1/2} E_0 \sin\frac{\omega}{c}(z - ct)$$

8.1 (a) $p_x = -qa, Q_{xx} = -3qa^2$
(b) $4\pi\epsilon_0 V = -q/4a$ at $x = 3a$, $-q/60a$ at $x = 9a$
(c) $4\pi\epsilon_0 V = -q/9a - q/9a = -2q/9a$ at $x = 3a$, $-q/81a - q/243a = -4q/243a$ at $x = 9a$

At $x = 3a$, the contributions of the dipole and quadrupole are equal, and together account for about 90 percent of the exact potential; at $x = 9a$, the contribution of the dipole is three times that of the quadrupole, and together they account for about $98\frac{1}{2}$ per cent of the exact potential.

8.2 $\mathbf{Q}' = \mathbf{Q} - 2\boldsymbol{p}\boldsymbol{r}_0 + q\boldsymbol{r}_0\boldsymbol{r}_0$

8.3 $4\pi\epsilon_0 V = qa^2(3\cos^2\theta - 1)/r^3$, where θ is the angle between $\boldsymbol{r}$ and the z-axis; see also Example 1.8. Note that in applying eq. (1.27) the scalar product for each dipole requires the cosine of an angle different from θ.

8.4 $p = 10^{-34}$ C m; much less.

8.5 See the figure below; $\alpha(\omega) \to \infty$ as $\omega \to \omega_0$ from below and $\alpha(\omega) \to -\infty$ as $\omega \to \omega_0$ from above.

8.6 See the derivation of eq.(8.43).

8.7 $\text{Re}\,\alpha(\omega) = (Q^2/m)(\omega_0^2 - \omega^2)/[(\omega_0^2 - \omega^2)^2 + \gamma^2]$
See the figure below; note the suppression of the divergence, as in Fig. 9.1.

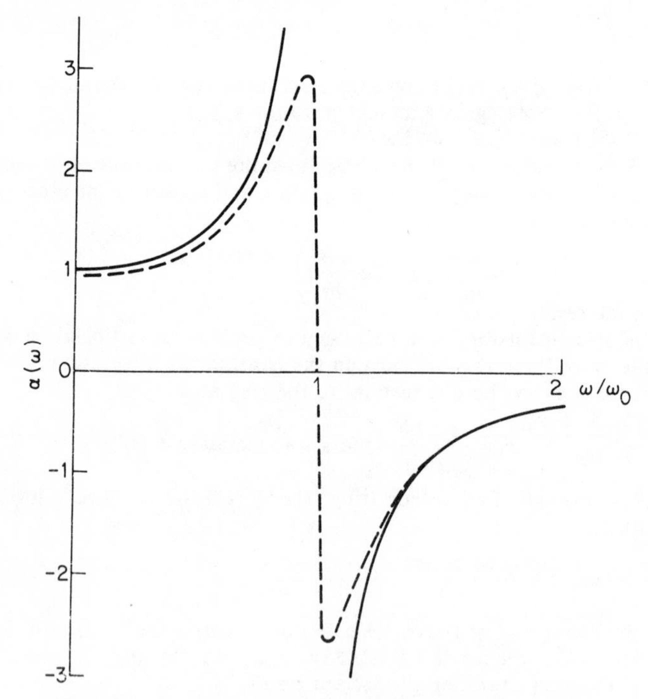

Fig. Ans.8 Frequency-dependent polarizability without damping (solid line), and real part of the frequency-dependent polarizability with damping (broken line)

9.1 1.00138

9.2 $\sigma = P\cos\theta$; $dq = \sigma da = \sigma r^2 \sin\theta\, d\theta$; $dE_p = dq/4\pi\epsilon_0 r^2 \to E_p$ on integration over θ.

9.3 1.371

9.4 $\coth u = 1/u + u/3 + \ldots$

9.5 1.21×10^{-3}

9.6 $e^{ux} \approx 1 + ux$ suffices

9.7 5.11×10^{-30} C m

9.8 0.336×10^{-40} F m^2

9.9 -20.2 deg mol^{-1} dm^2; -4.54×10^{-52} radian F m^3.

10.1 $\sqrt{(3)}\mu_B = 1.61 \times 10^{-23}$ J T^{-1}.

10.2 $\Delta W = 1.85 \times 10^{-24}$ J, $\nu = 2.80$ GHz, $T = 0.134$ K.

10.3 Use eq.(10.13).

10.4 $W = -\frac{1}{2}B^2[\chi V + \chi_0(V_0 - V)]/\mu_0$. Now $V = Ax$, where x = length of cylinder between poles, so $F = -\mathrm{d}W/\mathrm{d}x = \frac{1}{2}(\chi - \chi_0)AB^2/\mu_0 = \Delta\, mg$.

10.5 $2.14 \times 10^{-28}\ \mathrm{m^4\,H^{-1}}$ (or $\mathrm{A\,m^2\,T^{-1}}$)

10.6 $m_0 = 1.78 \times 10^{-23}\ \mathrm{A\,m^2} = 1.92\ \mu_B$. For $S = \frac{1}{2}$, $2[S(S+1)]^{1/2} = 1.73$.

12.1 $n(\omega) = \omega^2/\pi^2c^3$; $\varrho(\omega,T) = \hbar\omega^3/\pi^2c^3(e^{\hbar\omega/kT} - 1)$; $B_{lu}(\omega) = B_{lu}(\nu) \times 2\pi$.

12.2 $n(\sigma) = 8\pi\sigma^2$; $\varrho(\sigma,T) = 8\pi hc\sigma^3/(e^{hc\sigma/kT} - 1)$; $B_{lu}(\sigma) = B_{lu}(\nu)/c$.

12.3 $n(\lambda) = 8\pi/\lambda^4$ (ignore the negative sign in $\mathrm{d}N/\mathrm{d}\lambda$, which shows that λ increases as ν and ω increase).

12.4 $\lambda_{max}T = hc/5k(1 - e^{-hc/k\lambda_{max}T}) \approx hc/5k$. A more exact solution is $\lambda_{max}T = 0.2014 \ldots hc/k$.

12.5 $u = (8\pi k^4T^4/c^3h^3)\pi^4/15$; $R = 5.67 \times 10^{-8}\ \mathrm{W\,m^{-2}\,K^{-4}}$.

12.6 Set $\mathrm{d}N_u/\mathrm{d}t = \mathrm{d}N_l/\mathrm{d}t$ and express N_u and N_l in terms of N and ΔN to find $\Delta N/N = 1/(2B\varrho/A + 1) = 1/[2N(\nu,T) + 1]$. This is verified by setting $N_u = N_l\, e^{-h\nu/kT}$ to evaluate $\Delta N/N$.

12.7 Eqs (12.35) and (12.36) take the same form with σ replacing ν, but eq.(12.34) has $c\sigma$ replacing ν. Then $\mathrm{d}I/I = -BNh\sigma/S\Delta\sigma$ and the result follows.

14.1 Use $V_{1n} = \langle 1 | \hat{V} | n \rangle$

$$= (2qFa/\pi^2)\int_0^\pi y \sin y \sin ny\, \mathrm{d}y,$$

where $y = \pi x/a$, and set

$$\sin y \sin ny = \tfrac{1}{2}[\cos(n-1)y - \cos(n+1)y].$$

Then $V_{11} = \frac{1}{2}qFa$ as required, and for $n > 0$

$$V_{1n} = 0, \qquad n = 3,5,\ldots$$
$$= 8nqFa/\pi^2(n^2 - 1)^2,\ n \text{ even}.$$

(Note that the non-zero first-order correction implies a non-zero dipole moment for the system; as the system has a net charge, the moment depends on the choice of origin, as noted in Section 8.1. Fixing a charge $-q$ at $x = \frac{1}{2}a$ clearly gives a zero dipole moment and causes the first-order correction to vanish.)

14.2 The 2*s* wavefunction is spherically symmetric, while $2p_x$, $2p_y$ and $2p_z$ vary like x, y and z respectively. Then all matrix elements of z except that between 2*s* and $2p_z$ change sign on either side of one or more coordinate planes and so must be zero. Since $\langle 2s | \hat{V} | 2p_z \rangle = 3eFa = \langle 2p_z | \hat{V} | 2s \rangle$, then from eq. (14.35) the two perturbation energies are $\pm 3eFa$, corresponding to $a_{2s,2s} = 1/2^{1/2} = \pm a_{2s,2pz}$, or

$$| v_\pm \rangle = [| 2s \rangle \pm | 2p_z \rangle]/2^{1/2},$$

which are the usual *sp* hybrids.

14.3 Set $\Delta = 0$ and add or subtract the equations. Then $\mathrm{d}(c_1 \pm c_2)/\mathrm{d}t = \mp i\,(c_1 \pm c_2)V$ and the result follows on solving with the appropriate boundary conditions and substituting in eq.(14.102).

15.1 Use eq.(15.27) with $\omega_{mk} = 2\pi c/\lambda$ and $\omega = 0$; then $\alpha(0) = 3.18 \times 10^{-40}\ \mathrm{F\,m^2}$.

15.2 For the static ϵ_r, use $\alpha(0)$ from 15.1 in eq.(9.14); $\epsilon_r = 3.38$. From eq.(15.27), α is larger by 9/8 at 600 nm and by 4/3 at 400 nm. Then $n = 1.99$ at 600 nm and 2.31 at 400 nm.

15.3 In $\langle 0 | \hat{V} | 0 \rangle$ and $\langle 1 | \hat{V} | 1 \rangle$ the integrand is odd so that the integral is even. For $\langle 0 | \hat{V} | 1 \rangle$ integrate by parts using $\mathrm{d}(e^{-\xi^2}) = -2\xi e^{-\xi^2}\,\mathrm{d}\xi$ to obtain $-qE(\hbar/2m\omega_0)^{1/2}$. Then $W_0^{(2)} = -q^2/2m\omega_0^2 = -\frac{1}{2}\alpha E^2$, i.e. $\alpha = q^2/2m\omega_0^2$.

15.4 $$\hat{H} = \frac{-\hbar^2}{2m}\frac{\mathrm{d}^2}{\mathrm{d}y^2} + \frac{1}{2}m\omega_0{}^2y^2 - \frac{q^2E^2}{2m\omega_0^2}$$

Energy shifted by $-q^2E^2/2m\omega_0^2 = -\frac{1}{2}\alpha E^2$ as in Example 15.3, so that $\alpha = q^2/2m\omega_0^2$ again.

15.5 $W_0^{(2)} = -(e^2E^2m/2\pi\hbar^2a^2)\langle z^2(r+2a)e^{-2r/a}\rangle$. Either write $z = r\cos\theta$ and integrate in spherical polar coordinates, or note that the integrals in x^2 and y^2 have the same value as that in z^2 so that each is equivalent to that in $r^2/3$, and use the element of volume $4\pi r^2\,dr$. Then $W_0^{(2)} = -9me^2a^4E^2/4\hbar^2 = -9\pi\epsilon_0 a^3E^2 = -\frac{1}{2}\alpha E^2$, so that $\alpha = 18\pi\epsilon_0 a^3$ as required.

15.6 $\alpha_{SI} = 0.742\times10^{-40}\,\mathrm{F\,m^2}$; $\alpha_{cgs} = \alpha_{SI}/4\pi\epsilon_0 = 0.667\times10^{-24}\,\mathrm{cm^3}$; $\alpha_{au} = \alpha_{cgs}/a^3 = 4.5$.

15.7 Contribution from $|2p_z\rangle$ to $\psi_0^{(2)}$ is $-(2^{20}/3^{11})\,\pi\epsilon_0 a^3E^2 = -5.92\,\pi\epsilon_0 a^3E^2$. The total from $|np_z\rangle$ states is thus $-7.22\times\pi\epsilon_0 a^3E^2$, which is 80 per cent of the exact answer. The remainder comes from excitations from $|1s\rangle$ to the continuum of unbound states.

15.8 $\chi_d = -e^2a^2/2m$

16.1 1 at $r = 0$, 0.99921 at $r = 1$ nm.

16.3 $8.89\times10^7\,\mathrm{s^{-1}}$; invert for lifetime.

16.4 The scattered wavelengths are $1/\lambda_0 \pm 1/\lambda$, frequencies $c\lambda\lambda_0/(\lambda \pm \lambda_0)$.

Index